PYTHON PLAYGROUND: CODING GAMES AND PROJECTS FOR KIDS AND BEGINNERS

L.D. KNOWINGS

CONTENTS

INTRODUCTION

UNDERSTANDING THE EARLY STRUGGLES: MY CODING ODYSSEY

A few years ago, I stared at a screen filled with cryptic lines of code, feeling a mix of fascination and frustration. Maybe you've been there too – that moment when learning to code feels like deciphering an alien language, and each error message seems like a personal setback.

In those early days, it wasn't just about learning to code but overcoming the doubts and confusion accompanying a new challenge. Imagine sitting in front of a blank screen, questioning if you have what it takes to unravel the intricacies of Python. The initial wobbles and the fear of making mistakes were all part of my coding odyssey, and they might resonate with your journey.

I confess – a dance of mistakes marked my coding endeavors. From syntax errors that seemed to mock my every keystroke to logical blunders that left me scratching my head, the learning process was fraught with missteps. At times, it felt like I was taking two steps forward and one step back, caught in a choreography of errors.

There were moments of sheer frustration that made you question your decision to delve into the world of coding. Why couldn't my code behave as expected? Why did a seemingly simple task spiral into a cascade of errors? These questions echoed throughout my coding odyssey.

However, amidst the frustration, a turning point emerged. I began to view failure not as an obstacle but as a companion in my coding journey. Each mistake became a stepping stone, a lesson learned. The errors, rather than roadblocks, became signposts guiding me toward a deeper understanding of the coding.

In sharing these experiences, I aim to be transparent about my challenges. The unending trial and error loop was a significant part of my coding narrative. Yet, I navigated this loop and emerged as a more seasoned coder through perseverance, a willingness to learn from mistakes, and the acknowledgment that failure is not the end but a stepping stone.

THE QUEST FOR CLARITY: DECODING YOUNG MINDS:

In the bustling marketplace of coding resources, finding the perfect fit for young learners is akin to searching for a needle in a haystack. It's a challenging quest, where many offerings miss the mark entirely.

Some resources fall into the trap of oversimplification, leaving budding coders craving more substance and depth. It's like being served a tantalizing appetizer but never getting to savor the main course.

On the other hand, those resources drown learners in a sea of technical jargon and convoluted explanations. It's as if they're trying to scale a towering mountain with no ropes or safety nets, feeling overwhelmed and defeated by the sheer magnitude of the task.

In this digital jungle of coding resources, finding the elusive balance between simplicity and complexity, accessibility and challenge, can feel like an impossible feat that requires a guiding light to lead the way.

Despite these challenges, "Python Playground" offers a beacon of hope for aspiring coders. This book is not just another coding resource—it's a lifeline for those navigating the complexities of programming for the first time.

CONFRONTING COMPLEXITY WITH CONFIDENCE

Why did you decide to explore the world of coding? Was it a curiosity about technology, a thirst for a new challenge, or perhaps a practical need in our digital age? Your journey into coding has a catalyst, a unique spark. Understanding where you are in your life and what prompted this exploration is the key to tailoring our adventure to suit your needs.

"Python Playground" dismantles the walls of complexity brick by brick, providing a scaffolded approach to learning that builds confidence with each step.

Beginning with simple, easy-to-understand concepts and gradually progressing to more advanced topics, this book empowers learners to tackle complexity with courage and curiosity. Breaking down barriers and fostering a growth mindset will instill the confidence to conquer even the most challenging coding tasks.

A COMPANION TO DIGITAL CREATIVITY

Have you ever had a friend who knew all the best tricks and short-cuts in a game? Consider this book your coding companion, your knowledgeable friend in Python.

It shares insider tips, provides shortcuts, and guides you through the learning process with the expertise of a seasoned player. Learning Python becomes less of a solo mission and more of an interactive, enjoyable experience.

- Streamlined Learning: "Python Playground" offers shortcuts to streamline the learning process, helping you grasp Python concepts faster and more efficiently.
- Clear Understanding: Gain shortcuts to clear explanations, simplifying complex coding concepts for easy comprehension.
- Immediate Practice: Skip the theory and dive straight into practice with hands-on projects, gaining practical experience quickly.
- Creative Expression: Unleash your creativity with shortcuts to building games, animations, and websites, guided through each project.
- Engagement: Say goodbye to boredom with shortcuts to engaging projects and examples, making learning enjoyable and motivating.

Reading "Python Playground" isn't just about acquiring coding skills but finding a shortcut to a world of digital creativity. Three years ago, if someone had handed me a guide that turned coding puzzles into enjoyable activities, I would have leaped at the opportunity. That's precisely what this book offers – a path to creating your games, animations, and websites.

FROM LEARNER TO CREATOR

Fast-forward to today and I find myself not just as someone who learned to code but as a creator and architect of digital wonders. Picture yourself in this scenario—confident, proficient, and creatively expressing yourself through code.

With every project completed, a sense of accomplishment replaces the frustrating errors. That transformation is not just a possibility; it's the destination we aim for in the Python Playground.

Imagine not just understanding code but being able to create your digital wonders. This book takes you beyond theory, providing hands-on projects that turn coding into a canvas for your creativity.

- Mastery of Python Basics: By completing the projects and exercises in "Python Playground," you will achieve a solid understanding of fundamental Python concepts, including variables, loops, functions, and data structures.
- Practical Coding Skills: You will gain practical coding skills by applying Python to real-world projects and preparing for diverse coding challenges beyond the book.
- Foundation for Continued Learning: Armed with a strong understanding of Python and practical coding experience, you will be equipped to tackle more advanced topics and pursue further learning opportunities in dynamic programming.

You'll gain the practical skills to craft games, animations, and websites through playful projects and excellent examples. It's a hands-on approach to transforming coding from an abstract concept to a tangible, creative endeavor.

This guide is designed to be a light in the darkness, a companion that acknowledges your struggles and provides a clear path forward.

AUTHOR'S INSIGHT: A GUIDE WITH SHARED STRUGGLES

Why trust me as your guide through this coding journey? Because I've been in your shoes. I've faced the confusion, the uphill battles, and the triumphs of mastering coding. This book isn't just a compilation of information; it's a companion sharing experiences and insights gained from navigating the coding wilderness.

It's essential to recognize that even seasoned coders encounter setbacks, and the ability to learn and adapt from failures distinguishes a coder's journey. Through these experiences, I hope to convey the challenges and resilience that can turn the unending loop of trial and error into a path of continuous growth and success.

BRIDGING THE GAP: FROM THEORY TO PRACTICE IN PROGRAMMING

Before the availability of resources like "Python Playground," aspiring programmers often found themselves stuck in a frustrating cycle. They'd spend hours poring over textbooks and online tutorials, absorbing theoretical knowledge without clearly understanding how to apply it in real-world scenarios. The gap between theory and practice seemed overwhelming, leaving many disillusioned and discouraged.

Imagine searching for answers in a giant library where all the books are written in a language you don't understand, with no pictures or clear explanations to help you out. It was frustrating and confusing for many aspiring coders. They didn't know where to start or how to make progress, like trying to find their way in a dark forest without a flashlight.

Even if they figured out some bits and pieces, they still struggled to assemble it. It was like trying to build a house without a blueprint—things just didn't fit right. Plus, they had few opportunities to practice and improve their skills. It was like trying to learn to play a musical instrument without ever getting to touch one.

But now, with this book, learning to code is like having a clear roadmap. It's like having a friendly guide who speaks your language and shows you exactly what to do.

YOUR PERSONAL CODING COMPANION

With "Python Playground," you'll discover many engaging projects and interactive exercises that transform learning into a fun and rewarding experience. From building your games to creating dynamic animations, each activity is carefully crafted to ignite your curiosity and fuel your passion for coding.

A coding resource and immediately feeling a sense of connection, as if it were crafted just for you. That's the enchantment of "Python Playground" – a book that speaks directly to the hearts and minds of aspiring coders, no matter their level of expertise. It's like stumbling upon a treasure trove of coding adventures, just waiting to be explored.

But what truly sets "Python Playground" apart is its emphasis on practicality. Unlike other coding resources that leave you drowning in abstract theory, this book focuses on real-world applications, showing you how to turn your coding skills into tangible projects that showcase your creativity and ingenuity. It's like finding a roadmap to success explicitly tailored to your needs and aspirations.

So, if you've ever felt lost or overwhelmed on your coding journey, fear not. "Python Playground" is here to light the way, providing you with the guidance, support, and inspiration you need to thrive as a coder.

THE PYTHON ESSENTIALS

⬥

P sst, do you want to know a secret? Python isn't just a programming language—it's a magical portal to a world of endless possibilities and infinite creativity.

Here's the scoop for all you budding coders: Python isn't just your average programming language – it's the superhero of the coding world! Why? Well, buckle up because I'm about to spill the beans.

With Python as your trusty sidekick, you'll be conjuring up spells and weaving digital wonders faster than you can say "abracadabra"!

It's the late 1980s, and Guido van Rossum, a Dutch programmer, faces a dilemma – he needs a language that's powerful yet easy to learn and versatile yet simple to use.

And thus, in a stroke of genius, Python is born!!

Guido's brainchild wasn't just another programming language – it was a revolution waiting to happen. Inspired by his love for the British comedy show Monty Python's Flying Circus, Guido gave his creation a name that would forever cement its place in the annals of computer science.

INTRODUCTION TO PYTHON PROGRAMMING

Python is like the Swiss Army knife of coding languages. Need to build a website? Python's got you covered. Want to analyze mountains of data? Python's your best friend. From web development to artificial intelligence, Python can do it all – it's like having a superpower that adapts to whatever challenge you throw its way!

Its popularity stems from its accessibility and user-friendliness. With its straightforward syntax and focus on natural language, Python makes coding a breeze, allowing programmers to write and execute programs swiftly.

But here's the best part – Python isn't just for tech wizards and computer geeks. Nope, it's for everyone! Whether you're a seasoned programmer or a coding newbie, Python welcomes you with open arms.

Did I mention that Python is faster than a speeding bullet? Yep, you heard me right! Many developers agree that Python is more efficient, reliable, and lightning-fast than other programming languages. So when you're coding up a storm, Python's got your back – no cape required!

The Simplicity Saga: Python's Simple Syntax

Coding can be like learning a new language, but with Python, it's more like chatting with an old friend. The language's simple syntax and emphasis on natural language make it a breeze to learn and use. It's the Shakespeare of programming – eloquent, expressive, and loved by many.

Think of coding as cooking a meal. Just as a chef combines ingredients to create a dish, a programmer combines code to develop a program. Each ingredient (line of code) contributes to the overall flavor (functionality) of the dish (program). And just as a chef adjusts seasoning to taste, programmers fine-tune their code to achieve the desired outcome.

Let's talk about Python's three superpowers: readability, extensibility, and maintainability.

Readability: Python's syntax is so clear that it's like reading a well-written book. This makes it easy for you to understand your code and for others to join the coding party without feeling lost.

Extensibility: Python is like Lego for programmers. You have a small core language, like the basic Lego bricks, and a vast standard library – the extra pieces. Plus, you can easily add your pieces, creating a unique masterpiece.

Maintainability: Python follows the principle of "there should be one—and preferably only one—obvious way to do it." This not only makes coding standards happy but also keeps your projects readable and accessible to maintain

Young Coders Who Defied Expectations with Python

Alright, future coders, prepare to be inspired! Tell you about real-life superheroes who've used Python to create some seriously cool stuff.

Mark Zuckerberg

Mark Zuckerberg, the mind behind Facebook, didn't just become a tech titan overnight. Even in his early years, he recognized the value of learning to code. Mark's journey began in the 90s when he mastered BASIC programming as a kid.

His father enlisted software developer David Newman to provide private coding lessons at home, recognizing his potential. As he entered middle school, Mark was already proficient with computers, and coding was a hobby.

His early creations even made it into magazines. Mark Zuckerberg's story illustrates that with dedication and access to resources, coding can be a passion pursued alongside other interests, like his role as captain of his prep school fencing team.

Larry Page

Larry Page, co-founder of the ubiquitous Google, had a childhood surrounded by computers and science magazines.

His interest in coding ignited at an early age. Larry's diverse interests extended beyond programming; he was an avid reader and deeply appreciated music.

He attributed his musical education to the speed and efficiency later reflected in the development of Google.

Larry Page's legacy teaches us that a well-rounded set of interests, including coding, can shape a child's future and contribute to building innovative legacies.

Daphne Koller

Daphne Koller's journey challenges gender biases in programming. Today, Daphne, a professor of computer science at Stanford, faces a world where women in programming are rarely seen.

Undeterred, she forged her path, starting her coding journey at a young age. Her persistence and continuous improvement led her to her current esteemed position.

Beyond teaching, Daphne co-founded Coursera, an online platform offering programming courses curated by top institutions like Stanford. Her story inspires the breaking of barriers and underscores the importance of fostering diverse interests, especially for women in the tech world.

Kautilya Katariya

In the UK, Kautilya Katariya made headlines at six years old by becoming the youngest Guinness World Record Holder for AI programming. His secret? Completing a series of computer lessons from IBM. Kautilya began with independent study, diving into Python and IBM course materials.

Having completed several courses, including "Foundations Of AI," he showcases an exceptional dedication to learning. Kautilya believes that "cool things are run by a computer programmer or made using programming."

His passion, fueled by supportive parents and his drive, positions him to thrive in the future as he delves into advanced AI programming concepts.

SETTING UP AND WRITING YOUR FIRST SCRIPT

How To Install Python on Windows

Installing Python on Windows is a straightforward process. Follow these instructions to get Python up and running on your Windows computer:

1. Download the Python Installer:

- Navigate to the Python website.
- Click on the "Downloads" tab.
- Choose the latest version of Python for Windows and download the installer.

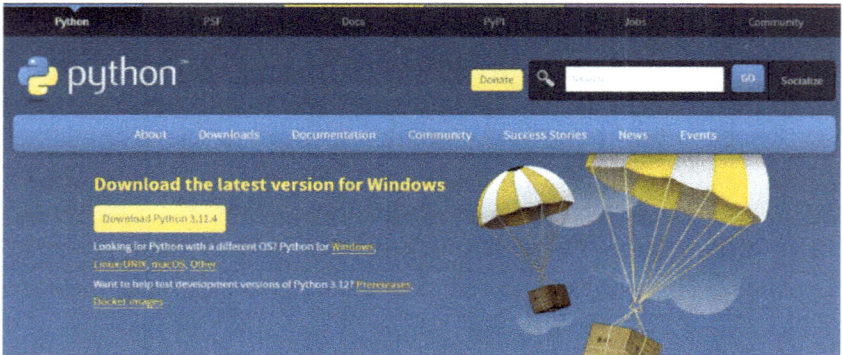

Python version	Maintenance status	First released	End of support	Release schedule
3.13	prerelease	2024-10-01 (planned)	2029-10	PEP 719
3.12	bugfix	2023-10-02	2028-10	PEP 693
3.11	bugfix	2022-10-24	2027-10	PEP 664
3.10	security	2021-10-04	2026-10	PEP 619
3.9	security	2020-10-05	2025-10	PEP 596
3.8	security	2019-10-14	2024-10	PEP 569

2. Run the Installer:

- Once the installer is downloaded, locate the downloaded file (typically in your Downloads folder) and double-click on it to run the installer.

3. Customize the installation (optional):

- The installer will open a setup window where you can customize your installation preferences.
- You can choose the installation directory, add Python to the system PATH, and select optional installation features.
- For beginners, sticking with the default settings is recommended, but feel free to customize as needed.

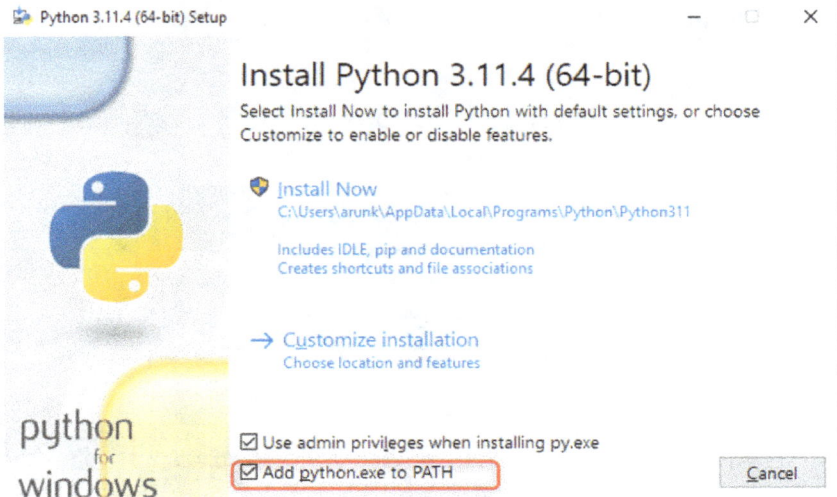

- At this point, choose the features you want to install. If all features are necessary, pick them and then click Next.

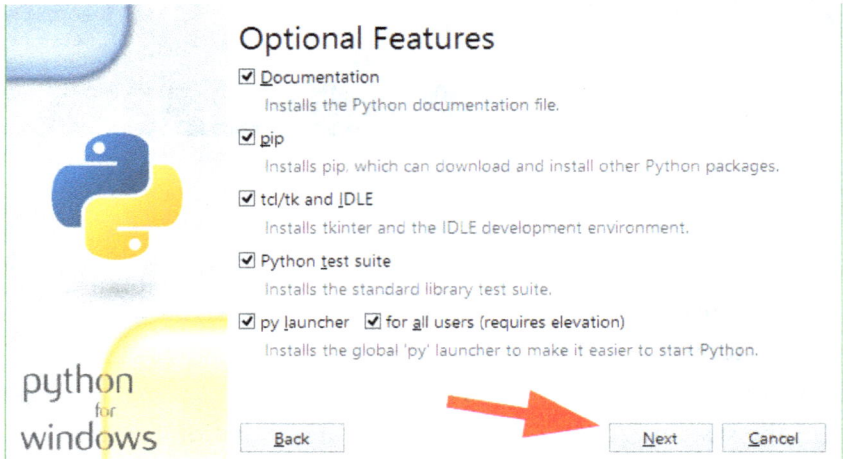

Optional Features

☑ Documentation
   Installs the Python documentation file.

☑ pip
   Installs pip, which can download and install other Python packages.

☑ tcl/tk and IDLE
   Installs tkinter and the IDLE development environment.

☑ Python test suite
   Installs the standard library test suite.

☑ py launcher ☑ for all users (requires elevation)
   Installs the global 'py' launcher to make it easier to start Python.

python for windows

Back   Next   Cancel

- An advanced options list will appear. Select the choices you need or all of them. Then click the Install button to start the process.

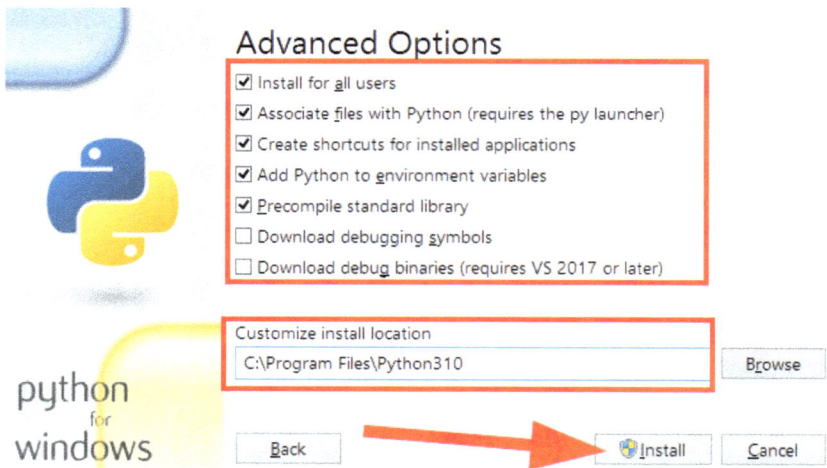

Advanced Options

☑ Install for all users
☑ Associate files with Python (requires the py launcher)
☑ Create shortcuts for installed applications
☑ Add Python to environment variables
☑ Precompile standard library
☐ Download debugging symbols
☐ Download debug binaries (requires VS 2017 or later)

Customize install location

C:\Program Files\Python310   Browse

python for windows

Back   🛡Install   Cancel

4. Install Python:

- After customizing your installation preferences, click the "Install Now" button to proceed.
- The installer will start installing Python on your system. This process may take a few minutes to complete.

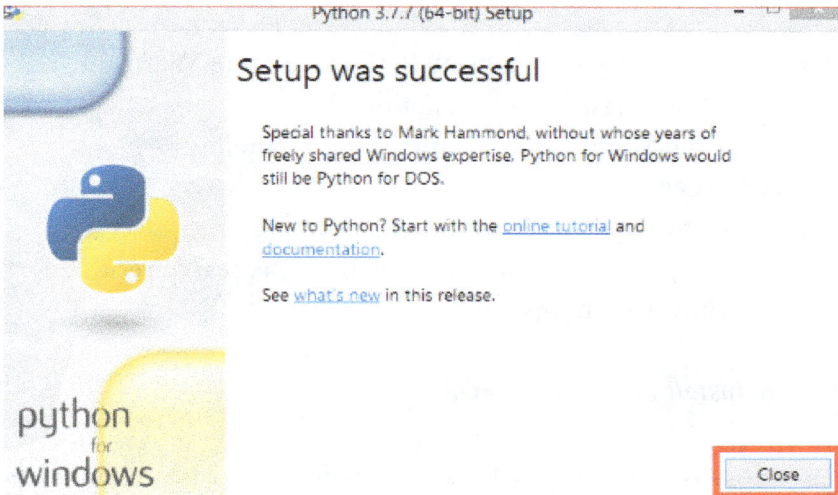

5. Verify the installation:

To make sure Python was installed correctly, do the following:

- Open PowerShell or the Command Prompt.
- Type python –version and press Enter in the Command Prompt window.
- 3. If you see a Python version, it indicates that Python has been successfully installed on Windows.

6. Alternate installation via Microsoft Store:

- Alternatively, you can install Python from the Microsoft Store for a hassle-free installation.
- Open the Microsoft Store app on your Windows computer.
- Search for "Python" in the search bar.
- Select the desired Python version from the search results and click "Get" to install it.

How to Install Python on macOS

Installing Python on macOS is a straightforward process. Follow these steps to get Python up and running on your Mac:

1. Check Python Version:

- Open Terminal by searching for it in Spotlight or navigating to Applications > Utilities > Terminal.
- Type the following command and press Enter:

```
python --version
```

- This command checks if Python is installed on your Mac and displays the installed Python version. If Python is not installed, you'll see a message indicating that Python is not found.

2. Visit the Python Website:

- Open your web browser and go to the official Python website.
- This is where you can download the macOS installer for Python.

3. Download the macOS Installer:

- On the Python website, click the "Download Python" button to download the macOS installer.
- Make sure to download the latest version of Python that is compatible with macOS.

4. Run the Installer and Follow the Instructions:

- Once the installer is downloaded, locate the downloaded file (usually in your Downloads folder) and double-click it to run the installer.
- Follow the on-screen instructions provided by the installer to complete the installation process.
- You may need to enter your macOS user password to authorize the installation.

Welcome to the Python Installer

This package will install **Python 3.10.1** for **macOS 10.9 or later**.

● **Introduction**

Read Me

Licence

Python for macOS consists of the Python programming language interpreter and its batteries-included standard library to allow easy access to macOS features. It also includes the Python integrated development environment, **IDLE**. You can also use the included **pip** to download and install third-party packages from the Python Package Index.

At the end of this install, click on Install Certificates to install a set of current SSL root certificates.

Go Back   Continue

5. Verify Python and IDLE are installed correctly:

- After installation, a desktop folder will appear. Click IDLE in this folder.

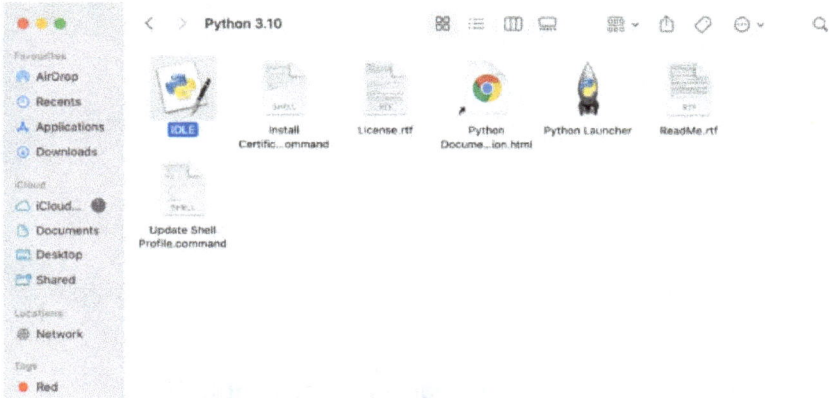

- Let's ensure that the most recent versions of Python and IDLE were set up correctly. To do that, double-click IDLE, the tool that comes with Python for combined development. If everything is okay, IDLE will show the Python shell like this:

● ● ●   **IDLE Shell 3.10.1**

```
Python 3.10.1 (v3.10.1:2cd268a3a9, Dec  6 2021, 14:28:59) [Clang 13.0.0 (clang-1
300.0.29.3)] on darwin
Type "help", "copyright", "credits" or "license()" for more information.
>>> |
```

- Write some basic Python code and execute it in IDLE. Enter the following and press return.

```
print("Hello, World!")
```

● ● ●   **IDLE Shell 3.10.1**

```
Python 3.10.1 (v3.10.1:2cd268a3a9, Dec  6 2021, 14:28:59) [Clang 13.0.0 (clang-1
300.0.29.3)] on darwin
Type "help", "copyright", "credits" or "license()" for more information.
>>> print("Hello, World!")
Hello, World!
>>> |
```

6. Verify installation with Terminal:

- The Terminal is another way to check whether the installation is complete. Enter this command in Terminal:

```
python3 --version
```

- Press Enter to view your newly installed Python version. This means that Python was successfully installed on your Mac.

How To Install Python on Linux

Installing Python on Linux is a relatively simple process. Follow these steps to get Python installed on your Linux system:

1. Check for Pre-Installed Python:

- Open a terminal window.
- Type the following command and press Enter:

```css
python --version
```

- This command will check if Python is installed on your Linux system and display the installed Python version. If Python is not installed, you'll see a message indicating that Python is not found.

2. Install via Package Manager:

- Most Linux distributions come with Python pre-installed. However, if Python is not installed or you need a different version, you can use your distribution's package manager to install it.
- For Debian-based systems (e.g., Ubuntu), use the following command:

```sql
sudo apt-get update
sudo apt-get install python3
```

- For Red Hat-based systems (e.g., CentOS), use the following command:

```
sudo yum install python3
```

3. Download the Latest Version of Python:

- If the version of Python available in your distribution's repositories is not the one you want, you can download and install the latest version from the official Python website.
- Visit the Python website at https://www.python.org/downloads/linux/ to download the latest version of Python for Linux.

4. Compile from Source (Optional):

- If you prefer to compile Python from source, download the source tarball from the Python website.
- Extract the tarball and follow the instructions in the README file to configure, compile, and install Python.

5. Verify Installation:

- After installation, you can verify that Python is installed correctly by opening a terminal window and typing the following command:

```css
python --version
```

- This command should display the installed Python version.

Voila, that's it! Now, you're ready to start writing and running Python scripts.

Writing and Running Your First Python Script

Now that you have Python installed, let's create a simple script and run it.

- Use a text editor like VSCode, Sublime Text, or even Notepad on Windows.

- In your text editor, type the following simple Python script:

```python
print("Hello, Python!")
```

- Save the file with a .py **extension, for example,** first_script.py.

Running Your Python Script

For Windows:

- Press "**Win + R,**" type cmd, and press Enter.
- Use the "**cd**" command to go to the directory where your script is saved.
- Type '**python first_script.py**" and press Enter.

For macOS/Linux:

- Use Spotlight on macOS or press **Ctrl + Alt + T** on Linux to open Terminal.
- Use the **cd** command to go to the directory where your script is saved.
- Type **python3 first_script.py** and press Enter.

Verify Output:

If everything is set up correctly, you will see the output:

```plaintext
Hello, Python!
```

You've just written and run your first Python script. Get ready to explore more.

THE MAGICAL MOMENT: WITNESSING YOUR CODE COME TO LIFE

In today's world, not everyone is fluent in programming. But if you've got the knack for it, you're holding a golden ticket. With information technology ruling the roost, we're living in a time when being able to whip up your digital masterpiece opens up a world of possibilities.

Coding: Create Something from Nothing

Coding is not just about punching lines of text into a computer. It's an art form, a creative outlet where you sculpt something from nothing. Sure, you see mobile apps and websites popping up left, right, and center, but have you ever wondered how they work? Believe it or not, for many, it remains a mystery.

Let's set the stage for that exhilarating moment when your code bursts into life for the first time.

Picture the electrifying moment when your code springs to life for the first time. After investing hours of brainpower, countless lines of code, and a few frustrated sighs. You've meticulously crafted every function and debugged every glitch, and now, as you hit that "run" button, there's this moment of breathless anticipation.

And then, it happens.

Your screen lights up with the fruits of your efforts. Maybe it's a sleek interface for a mobile app, or perhaps it's a beautifully responsive website. Whatever it is, it's alive, it's working, and it's all thanks to you. It's like witnessing a digital birth—an exhilarating rush of adrenaline mixed with a deep sense of satisfaction.

That moment is more than just lines of code coming together. It's a validation of your skills, a testament to your perseverance, and a glimpse into the endless possibilities of the digital world. It's the moment when all the late nights and caffeine-fueled coding sessions pay off in the most glorious way possible.

So, when I talk about the excitement of coding, I'm talking about moments like these. Seeing your creations come to life is addictive and empowering, and it keeps me coming back for more.

A Reminder!

Aim high; you will still be among the stars if you fail.

See, in this coding world, there's a hierarchy—the rock stars, the coding ninjas who make it all seem like a piece of cake. Then, there's the rest of us, firmly rooted in the real world of debugging and trial and error.

Here's the deal: Cut yourself some slack. Don't set the bar too high; celebrate every small victory, and embrace that coding is complicated—perfectly okay.

UNDERSTANDING DATA TYPES AND BASIC OPERATIONS:

Data types are like code superheroes, each with distinct skills. Standard or built-in Python data types are as follows:

Numeric:

This includes integer (whole numbers) and float (decimal numbers) data types.

Example: Imagine you're managing your finances. The money in your bank account would be an integer (like $1000), while the amount you spent on groceries might be a float (like $42.56).

Sequence Type:

These are ordered collections of similar or different data types.

Example: Think of a shopping list. Each item on the list is like an element in a sequence. Whether it's a list of groceries or a playlist of songs, sequence types keep things organized in a specific order.

Boolean:

This data type has only two possible values: True or False.

Example: Imagine a traffic light – it's either red or green. Similarly, boolean types in coding are all about binary choices – True or False. They help us make decisions in our programs.

Set:

Sets are unordered collections of unique elements.

Example: Think of a collection of unique keys. Each key in a keychain is like an element in a set. Even if you have multiple copies of a key, the set only keeps one of each unique key.

Dictionary:

Dictionaries are collections of key-value pairs, with unique keys accessing each value.

Example: Imagine a dictionary as a phone book. The names are like keys, and the corresponding phone numbers are the values. When you look up a name (key), you get the associated phone number (value).

Binary Types (memory view, bytearray, bytes):

These data types deal with binary data and are commonly used for low-level manipulation.

Example: When you take a photo with your phone, the image is stored in binary format. Memoryview, bytearray, and bytes allow programmers to work directly with this binary data, performing tasks like image processing or file manipulation.

String:

Strings are sequences of characters typically used for text.

Example: A person's name, like "John" or "Alice," is a string. It could also be a sentence like "Hello, how are you today?"

1. Arithmetic Operators:

- **Addition** (+): Picture adding slices of pizza together to make a delicious pie!
- **Subtraction** (-): Imagine removing toppings from a pizza to customize it as you like.
- **Multiplication** (*): It's like arranging pizza boxes in rows and columns to see how many you have.
- **Division** (/): Think of sharing a pizza equally among friends and figuring out how many slices each person gets.

Example: Feeling like a math whiz? Let's add some numbers together! Try this:

```python
num1 = 10
num2 = 5
result = num1 + num2
print("The sum is:", result)
```

2. Comparison (Relational) Operators:

- Greater Than (>): Picture two stacks of pizza boxes, and you're deciding which one is taller.
- Less Than (<): Imagine comparing the height of two animals to see which is shorter.
- Equal To (==): It's like checking if two pizzas have the same number of slices.
- Not Equal To (!=): Compare toppings on two pizzas to see if they differ.

Example: Want to compare some values? Let's see if ten is greater than 5:

```python
print(10 > 5)  # True
```

3. Assignment Operators:

- Assign (=): Hand out pizza slices to each party member.
- Addition Assignment (+=): Adding extra toppings to your pizza slice.

- Subtraction Assignment (-=): Removing toppings from your pizza slice.
- Multiplication Assignment (*=): Doubling the number of pizza slices you have.

Example: Time to assign some values! Let's give a variable a new value.

```python
x = 5
x += 3   # Equivalent to x = x + 3
print("New value of x:", x)   # Should be 8
```

4. Logical Operators:

- AND (and): Imagine deciding to go to the beach only if it's sunny AND warm.
- OR (or): It's like planning a picnic and being happy to go if it's sunny OR not too windy.
- NOT (not): Picture not swimming because the water is NOT warm.

Example: Let's play with logic! Are both conditions true?

```python
sunny = True
warm = True
print(sunny and warm)   # True
```

5. Bitwise Operators:

- AND (&): Think of overlapping areas between two Venn diagrams.
- OR (|): Imagine combining two sets of puzzle pieces to see the complete picture.
- XOR (^): Picture selecting puzzle pieces from two different puzzles and combining them in a new, unique way.

Example: Dive into the world of bits! Let's perform a bitwise AND operation.

```python
num1 = 10    # 1010 in binary
num2 = 7     # 0111 in binary
result = num1 & num2
print("Bitwise AND result:", result)   # Should be 2
```

6. Membership Operators:

- In: Imagine checking if an apple is in the fruit list.
- Not In It's like seeing if a particular fruit is not on the fruit list.

Example: Let's check if 'apple' is in the list:

```python
fruits = ['apple', 'banana', 'orange']
print('apple' in fruits)   # True
```

7. Identity Operators:

- Is: Picture two pizzas and check if they are the same size and have the same toppings.
- Is Not: It's like comparing two different pizzas and confirming they're not identical.

Example: Are two variables the same thing? Let's find out:

```python
x = [1, 2, 3]
y = [1, 2, 3]
print(x is y)   # False, because they refer to different objects
```

INTERACTIVE ELEMENT

Python Basics Quiz

Let's test your knowledge of Python basics! Answer the following questions to see how much you've learned:

1. What is the output of the following code?

```python
print(2 + 3 * 4)
```

    a) 20
    b) 14
    c) 15
    d) 24

Answer: b) 14

2) What will be the result of the following code snippet?

```python
x = 10
y = 5
print(x > y)
```

a) True
b) False
c) Error
d) None

Answer: a) True

3) Which is the correct way to declare a variable in Python?

a) var x = 5
b) x = 5
c) int x = 5
d) variable x = 5

Answer: b) x = 5

4) What does the '+= 'operator do in Python?

a) Adds two numbers
b) Subtracts two numbers
c) Multiplies two numbers
d) Adds the right operand to the left operand and assigns
the result to the left operand

Answer: d) Adds the right operand to the left operand and assigns the result to the left operand

5) How do you create a comment in Python?

a) //This is a comment
b) #This is a comment
c) /This is a comment/
d) <!--This is a comment-->

Answer: b) #This is a comment

6) Which of the following is NOT a valid data type in Python?

a) int
b) float
c) char
d) str

Answer: c) char

7) What is the correct way to create a function in Python?

a) function myFunction():
b) def myFunction():
c) create myFunction():
d) func myFunction():

Answer: b) def myFunction():

8) What will be the output of the following code?

```python
x = "Hello"
print(x[2:])
```

a) lo
b) llo
c) Hell
d) Hello

Answer: a) lo

9) How do you print "Hello, World!" in Python?

a) print("Hello, World!")
b) console.log("Hello, World!")
c) echo "Hello, World!"
d) printf("Hello, World!")

Answer: a) print("Hello, World!")

10) What does the 'input()' function do in Python?

- a) Displays output to the console
- b) Reads input from the user
- c) Converts a string to an integer
- d) Checks if a variable is defined

Answer: b) Reads input from the user

Once you've answered all the questions, check your answers. Good luck!

SEGUE

In this chapter, we embarked on an exciting journey into Python programming fundamentals.

Key Takeaways:

- Python's simplicity, readability, and flexibility make it an ideal language for beginners and coding experts.
- Real-life examples of young coders breaking barriers with Python highlight its accessibility and empowerment for all demographics.
- The installation guide ensures that users of all operating systems can easily set up Python and start coding immediately.
- Learning about data types and operations in Python provides a comprehensive understanding of data manipulation and analysis capabilities.

So there you have it, young coder. It's time to translate these concepts into action! Experiment with diverse data types, explore various operations and push the boundaries of creativity with Python.

But wait, there's more! Now that you've mastered the basics, let's dive into data handling and visualization in the next chapter.

Get ready to unlock the power of organizing and presenting information in Python – where numbers and facts come to life through your code. Get set for an adventure into the world of data!

DATA HANDLING AND VISUALIZATION

H ey there, data explorer! Have you ever felt like you're drowning in a sea of numbers and spreadsheets, desperately searching for a lifeboat to rescue you from boredom? Well, get ready to toss those life jackets aside because we're about to embark on a thrilling adventure into the world of data handling and visualization using Python!

Say goodbye to dull data and hello to exciting graphs, charts, and visual stories that will make you the hero of your data-driven quests. Get ready to dive in because the waters of data have never been more exhilarating!

DATA MANIPULATION WITH PANDAS AND NUMPY

First, we're introducing you to the dynamic duo Pandas and NumPy.

Have you ever heard of Pandas? No, not the cute bears, but the Python library that makes handling tabular data as smooth as slicing through butter!

These are the superheroes of data manipulation in Python. They are your trusty sidekicks, helping you organize, filter, and transform your data. We're turning you into a data maestro in no time!

Pandas is like the wizard of the Python world for dealing with tables of data, especially the ones you find in Excel or those comma-separated .csv files. It's the superhero cape you need for data analysis because it's super user-friendly.

You might wonder, "What sets Pandas apart from other Python libraries?" Great question! While libraries like XLSX and XLRD handle Excel data behind the scenes, Pandas takes the lead in interacting with Excel spreadsheets and CSV files, making it the go-to choice for data analysis lovers like yourself.

But wait, what's a .csv file? It's like the VIP format in the tabular data club. Whether it's .csv, .tsv, or any other variation, it follows the same rules—a unique character waving the way through the table, guiding you from one cell to the next.

The Essence of a DataFrame

In Pandas' enchanting realm, a DataFrame is a pivotal player. According to the Pandas library documentation, a DataFrame is a "two-dimensional, size-mutable, potentially heterogeneous tabular data structure with labeled axes (rows and columns)." That might sound a bit complex at first, but fear not—let's break it down into beginner-friendly terms.

Think of It as a Table:

In plain language, a DataFrame is like a digital table for your data. It's a powerhouse that organizes information in a structured and meaningful way. Let's explore the key characteristics that make a DataFrame a game-changer:

1. Multiple Rows and Columns:

Imagine your data table as a chessboard with multiple rows and columns. Each square on this board can hold a unique piece of information, forming a rich tapestry of data.

2. Each Row, a Story to Tell:

Each row in your DataFrame is like a story waiting to be told. It represents a sample of data—a specific instance or observation contributing to the bigger picture.

3. Columns, the Variables:

Now, think of columns as the storytellers. Each column contains a different variable describing the samples represented by the rows. It's like having different chapters in your data story.

4. Uniform Data Types:

In this digital narrative, every column usually uses the same type of data. Harmonious consistency keeps your data organized, whether numbers, strings, or dates.

5. No Missing Value Drama:

Here's where Pandas outshines even Excel – DataFrames steer clear of missing value drama. Unlike Excel datasets with gaps and empty spaces, a DataFrame ensures a seamless and complete flow of information between rows and columns.

CRACKING THE CODE WITH NUMPY: YOUR MATH BUDDY IN PYTHON!

Meet your new best friend – NumPy! It's like having a math wizard by your side, making complex calculations as easy as pie. Let's unlock the magic together and discover how NumPy becomes your ultimate companion.

What's NumPy?

NumPy is like the rockstar of Python libraries. It stands for "Numerical Python," it's your go-to when you want to play with arrays, which are organized lists of numbers.

Arrays, Arrays, Arrays!

Imagine you have a list of your favorite numbers: [3, 6, 9, 12, 15]..
Now, you can turn this list into a powerful array with NumPy. Arrays make it super easy to perform math operations on your numbers.

```python
import numpy as np

numbers = np.array([3, 6, 9, 12, 15])
```

Let's Do Some Math!

Let's throw in some math operations like we did back in school.

- **Addition & Subtraction:**

```python
result_addition = numbers + 5
result_subtraction = numbers - 3
```

you added 5 to each number and subtracted 3. Easy, right?

- **Multiplication & Division:**

```python
result_multiplication = numbers * 2
result_division = numbers / 3
```

Now, you're multiplying each number by two and dividing by 3. It's just like handling pocket money!

Playing with More Advanced Math:

NumPy is not just about basic operations; it can handle more complex math, too.

- **Square Root:**

```python
result_sqrt = np.sqrt(numbers)
```

You just calculated the square root of each number. Fancy!

- **Sum and Average:**

```python
total_sum = np.sum(numbers)
average = np.mean(numbers)
```

Now, you've found your array's total sum and average –like acing your math tests!

VISUALIZING DATA USING MATPLOTLIB AND SEABORN:

Picture this: you've got a bunch of numbers sitting there, staring back at you from your screen. They're just... Well, numbers. But what if I told you that with Matplotlib, those numbers can come alive, morphing into beautiful graphs and charts that tell stories? Yep, that's right—Matplotlib is like the artist of the data world, turning dull digits into captivating visuals.

So, how does Matplotlib work its magic? Let me break it down for you:

1. Easy Peasy Plots:

First, Matplotlib is all about making things easy. It's like the friend who always has your back when you need to whip up a quick plot. You can create stunning visuals that bring your data to life with just a few lines of code.

Matplotlib is the Picasso of Python visualizations. It's not just about making things easy; it's about turning the seemingly impossible into a reality.

2. Crafting Visual Ensembles:

Think of Matplotlib as your data's stylist. From sleek line plots that exude sophistication to bar charts that make a bold statement, Matplotlib tailors your data's visual ensemble to match its personality.

3. Interactive Adventures:

But wait, there's more! Matplotlib doesn't just stop at static visuals. Nope, it's all about interactivity too. Imagine zooming in on a plot, panning around to explore different areas, or even dynamically updating the data—all with a few clicks or swipes. That's the kind of magic Matplotlib brings to the table.

4. Customization Galore:

Now, let's talk about personalization. With Matplotlib, you're not limited to cookie-cutter plots. Customize your visuals to match your personality – from pop color palettes to layouts that tell a story. Your data, your rules!

5. Exporting Adventures – Share Your Stories:

Matplotlib ensures your creations don't stay confined. Export your visual sagas to various file formats, making them ready to dazzle in reports, presentations, or even on your social media stage.

6. Seamless Integration – JupyterLab and Beyond:

Whether you're a JupyterLab enthusiast or navigating through GUIs, Matplotlib seamlessly integrates into your playground. Embed your creations effortlessly, making your data storytelling journey a breeze.

7. A Galaxy of Possibilities with Third-Party Packages:

Matplotlib isn't just about what it can do; it's about the possibilities it opens up. Dive into a rich array of third-party packages built on Matplotlib, each offering a unique twist to elevate your data narratives.

CHART CHARM WITH SEABORN: ELEVATING YOUR DATA AESTHETICS!

All right, folks, get ready to take your charts to the next level because Seaborn is about to make your data look more incredible!

So, do you know how charts and graphs are great for visualizing data? With Seaborn, they're not just great—they're downright awesome! This Python library is like a stylist for your data, adding a bit of flair that takes your visuals from "meh" to "wow!"

And the best part? Seaborn's got a knack for making things easy. Its plotting functions are like magic wands, waving away the complexities of data visualization so you can focus on what matters: the story behind the numbers.

With Seaborn, you can spice up your charts with vibrant colors, sleek styles, and eye-catching details. It's like giving your data a makeover and watching it shine like never before. Seaborn plays well with others, seamlessly integrating with pandas data structures and Matplotlib to make your plotting experience a breeze.

SEABORN IN ACTION: REAL-LIFE DATA VISUALIZATION ADVENTURES

Are you ready to witness the transformative power of data visualization in action? Brace yourself for a captivating journey as we delve into the real-life applications of Seaborn, the enchanting Python library that turns mundane data into visual masterpieces.

Distribution Plot for Customer Details

Picture this: you're working for a trendy fashion retailer, and your team wants to understand the age range of your most loyal customers. You've got a mountain of data at your fingertips, including the ages of shoppers who frequent your stores and website.

You turn to Seaborn and its trusty distribution plots to make sense of this data. These plots, also known as histograms, are like visual snapshots of the age distribution among your customers. Each bar in the histogram represents a different age group, and the height of the bar shows how many customers fall into that group.

As you examine the distribution plot, trends emerge. Maybe there's a spike in the number of shoppers in their twenties, indicating a strong presence among young adults. Or perhaps there's a broader distribution across age groups, suggesting a diverse customer base.

Whether it's launching new collections aimed at millennials or offering special promotions for older shoppers, Seaborn's distribution plots help you make informed decisions that resonate with your customers and keep them coming back for more.

Barplot for Football Player Activity:

Imagine sitting in a packed stadium, eagerly awaiting kickoff in a high-stakes football match, and the legendary Lionel Messi strides onto the pitch. You're not just a spectator but a keen observer, hungry for insights into Messi's mesmerizing performance. Here's where Seaborn steps up, offering a glimpse into Messi's brilliance with its dynamic bar plots.

Seaborn can craft captivating bar plots that unravel Messi's movements and actions throughout the game. Picture this: each bar represents a different facet of his gameplay—successful dribbles, accurate passes, shots on target, and goals scored. As you delve into the bar plot, you're treated to a visual feast of Messi's dominance on the field—whether he's weaving through defenders with his trademark dribbling, threading inch-perfect passes to teammates, or unleashing thunderous shots on goal.

With Seaborn's intuitive visuals, it's like having a front-row seat to Messi's magic, allowing you to marvel at his skill and creativity with every touch of the ball.

Boxplot for the effectiveness of treatment

Let's say you're a data analyst at a healthcare research institute studying the effectiveness of different treatments for a particular medical condition. You have a dataset containing the recovery times of patients who underwent various treatment options. To compare the distribution of recovery times across treatments, you decide to use box plots.

In this scenario, you can create a boxplot for each treatment group, where the box represents the interquartile range (IQR) of recovery times, a line inside the box indicates the median and the whiskers extend to show the range of the data. Outliers, if any, are represented as individual points beyond the whiskers.

By visualizing the distribution of recovery times using box plots, you can quickly identify which treatments result in faster recoveries and any significant variations or outliers that may warrant further investigation. This helps healthcare professionals make informed decisions about treatment options and improve patient outcomes.

INTERACTIVE VISUALIZATIONS WITH PLOTLY AND BOKEH

Hey there, future data wizards! Today, let's spill the tea on Plotly and Bokeh—the dynamic duo of data visualization! Forget everything you knew about charts because, unlike their buddies Matplotlib and Seaborn, Plotly and Bokeh are not just Python players; they're the HTML and JavaScript maestros who elevate your visualizations to a new level!!

Say Hello to Plotly

Plotly is like the magician of interactive graphs. Its Python magic wand can conjure plots that do more than sit there looking pretty.

What sets Plotly apart? Picture this: tooltips that spill the beans on intriguing markers. Hover over a point, and bam! Plotly reveals the hidden gems, turning your data into a treasure trove of insights. It's like having your data wizard right at your fingertips!

And here's the icing on the cake – Plotly allows you to capture the magic. Save your interactive chart as a PNG, preserving the dynamic elements that make your data come alive. Whether for presentations, reports, or sharing insights, Plotly ensures that your visualizations remain as captivating as the moment you first explored them

Behind the scenes, Plotly draws inspiration from the renowned d3.js JavaScript library, adding an extra layer of sophistication to its repertoire. Think of Plotly as your Python-friendly bridge to the world of stunning, interactive charts. It's not just a tool; it's an invitation to turn your data into a visual masterpiece, all within the comfort of Python's embrace.

Plotly Power: Where Data Becomes an Interactive Adventure

Get ready to discover a realm where charts come to life, insights unfold with a hover, and every click takes you deeper into the magic of your data story. This is Plotly – the wizard of interactivity, the storyteller of data. Let the adventure begin!

1. Interactive Wand Magic: Ever wished your charts could talk? Plotly makes it happen. With interactive tooltips, each data point becomes a storyteller. Hover over them, and voila! Insights pop up like magic. It's like having a personal guide through your data wonderland.

2. Zooming Extravaganza: Forget static views. Plotly gives you the power to zoom into the nitty-gritty details. Do you have a specific point that needs attention? Zoom in. Are you exploring a particular region's nuances? Zoom in. It's like having a data magnifying glass, revealing hidden gems in every pixel.

3. Dynamic Data Narratives: Plotly isn't just about charts; it's about creating dynamic data narratives. Your visualizations become stories that unfold as you explore. It's like turning your data into a page-turner novel – every click reveals a new chapter of insights.

4. Magic Saver—PNG Edition: Have you captured the perfect moment in your interactive chart? Plotly lets you save it as a PNG, freezing that dynamic vibe. Whether for presentations, reports, or sharing with the world, your visualizations remain as captivating as the first exploration.

Say Hello to Bokeh

And now, let's turn our attention to Bokeh – the master of HTML and JavaScript wizardry.

Bokeh struts its stuff as the rockstar of interactive data visualization libraries. Unlike its buddies Matplotlib and Seaborn in the Python visualization scene, Bokeh brings HTML and JavaScript to the party to render graphics. It's not just a web wizard, though – consider Bokeh your versatile sidekick, equally adept at unraveling data mysteries or creating chart masterpieces that will have your projects and reports stealing the spotlight!

But hold on tight because Bokeh isn't just reserved for the web wizards. It's also a powerhouse for anyone who wants to dive deep into their data or whip up some eye-catching charts.

Now, buckle up because we're about to take a joyride into the world of Bokeh! Imagine you have a school project or a report that needs some visual severe oomph – that's where Bokeh steps in like the superhero it is.

1. HTML and JavaScript – Bokeh's Secret Sauce:

So, here's the inside scoop. Bokeh doesn't just speak Python; it's bilingual! It flirts with HTML and JavaScript to give your charts that extra sparkle. It's like having a multilingual friend to chat with everyone at the party.

2. Bokeh for Every Data Explorer:

You might think, "Do I need to be a coding whiz to use Bokeh?" Not! Bokeh is like your friendly neighborhood superhero that welcomes everyone. Whether you're a curious data explorer or a student with a knack for turning numbers into visual poetry, Bokeh is your go-to sidekick.

3. Eye-Catching Charts for School Projects:

You're working on a school project about your favorite animals. Bokeh transforms your boring bar graphs into a visual zoo! Picture colorful, interactive charts that bring your data to life. Suddenly, your project isn't just informative; it's a journey into the wild world of your favorite creatures.

4. Bokeh's Plotting Playgrounds:

Bokeh isn't just about making charts; it's about creating entire playgrounds for your data. With Bokeh, you can whip up plots that tell stories. Want to showcase how temperatures change throughout the year? Bokeh has your back. It's like having a digital canvas where your data becomes a work of art.

5. The Bokeh Effect – More Than Just Charts:

But wait, there's more! Bokeh isn't confined to the world of charts alone. It's your creative partner in crime for creating interactive dashboards. Imagine having a dashboard that lets you explore data like a treasure hunt – clicking, zooming, and discovering insights like an actual data adventurer.

Ah, let me tell you about the incredible features that make Bokeh shine brighter than a shooting star in the night sky!

Interactive Magic: Bokeh isn't your average plotting library – it's a wizard that brings your data to life! With its interactive features, you can zoom, pan, and hover over your charts to explore every detail up close. It's like giving your audience a front-row seat to the data show!

Web-Friendly: Say goodbye to static plots and hello to web-based wonders! Bokeh is built for the digital age, using HTML and JavaScript to create stunning visualizations that shine bright on any screen. Whether you're building a dashboard or an app, Bokeh has your back with its web-friendly magic.

Versatile Plotting: Need a scatter plot? No problem! Want a bar chart? You got it! Bokeh offers a treasure trove of plotting options, from simple line graphs to complex heatmaps. Its versatile toolkit lets you turn your data into any desired chart.

Seamless Integration: Whether you're a Python pro or a newbie, Bokeh plays nice with all your favorite tools and libraries. It seamlessly integrates with Pandas, NumPy, and other Python powerhouses, making it easy to incorporate into your data science workflow.

Cross-Platform Compatibility: Whether you're on Windows, Mac, or Linux, Bokeh has got you covered. It's designed to work seamlessly across different platforms, so you can easily create and share your visualizations, no matter where or what device you're using.

INTERACTIVE ELEMENT

Data Quiz Challenge

Get ready for a fun and interactive challenge with our Data Quiz!

Instructions:

- Read each question carefully.
- Choose the most appropriate answer.
- Select your answer and move on to the next question

Question 1:

You have a dataset showing the sales performance of different products over the past year. Which type of graph would be best for comparing the sales of each product?

   a) Pie Chart
   b) Bar Chart
   c) Line Chart
   d) Scatter Plot

Answer: b) Bar Chart

Question 2:

Which type of graph would be most suitable for visualizing the distribution of ages in a population survey?

   a) Histogram
   b) Box Plot
   c) Line Chart
   d) Scatter Plot

Answer: a) Histogram

Question 3:

You're analyzing the correlation between temperature and ice cream sales. Which type of graph would help you visualize this relationship?

a) Line Chart
b) Scatter Plot
c) Bar Chart
d) Pie Chart

Answer: b) Scatter Plot

Question 4:

You're presenting the results of a survey on favorite pizza toppings. Each respondent selected multiple toppings. Which type of graph would effectively display this data?

a) Pie Chart
b) Bar Chart
c) Line Chart
d) Histogram

Answer: a) Pie Chart

Question 5:

You want to compare the performance of different machine learning models based on their accuracy scores. Which type of graph would be most appropriate?

a) Bar Chart
b) Line Chart

c) Box Plot

d) Scatter Plot

Answer: a) Bar Chart

Question 6:

You have a dataset showing the monthly rainfall in different cities. Which type of graph would be best for comparing the rainfall across cities?

a) Pie Chart

b) Bar Chart

c) Line Chart

d) Scatter Plot

Answer: b) Bar Chart

Question 7:

If you're analyzing the distribution of test scores in a class, which type of graph would be most suitable?

a) Box Plot

b) Histogram

c) Line Chart

d) Scatter Plot

Answer: b) Histogram

Question 8:

If you want to visualize the trend in stock prices over the past year, which type of graph would be most appropriate?

a) Pie Chart
b) Bar Chart
c) Line Chart
d) Scatter Plot

Answer: c) Line Chart

Question 9:

You're comparing the popularity of different social media platforms among teenagers. Which type of graph would effectively display this data?

a) Pie Chart
b) Bar Chart
c) Line Chart
d) Scatter Plot

Answer: b) Bar Chart

Question 10:

You have survey data on customer satisfaction ratings for a product. Which type of graph would be best for showing the distribution of ratings?

a) Pie Chart
b) Bar Chart
c) Box Plot
d) Scatter Plot

Answer: c) Box Plot

SEGUE

You've learned how to turn boring numbers into captivating stories through nifty charts and graphs using Python libraries like Matplotlib, Seaborn, Plotly, and Bokeh.

Key Takeaways:

- Visualizing data is essential for understanding trends, patterns, and relationships within datasets.
- Different graphs are suitable for different data types, such as bar charts for comparisons, histograms for distributions, and scatter plots for correlations.
- Libraries like Matplotlib, Seaborn, Plotly, and Bokeh offer powerful tools for creating interactive and visually appealing visualizations.

Now, it's time to put your newfound skills into action! Take the concepts you've learned in this chapter and start visualizing your datasets. Whether you're analyzing sports scores, survey results, or stock prices, let your creativity shine through in your visualizations.

In the next chapter, we will take your Python skills to the next level. Get ready to delve into the exciting world of Python applications and automation. You'll learn to build your apps and automate tedious tasks like a tech wizard! So, stay tuned and keep exploring Python's limitless possibilities.

Now, go forth and visualize your world with Python!

BUILDING AND AUTOMATING
WITH PYTHON

I magine this: You have an idea for a website you're passionate about. Perhaps it's a blog where you can share your thoughts, an online portfolio showcasing your creative projects, or even a tiny e-commerce site selling handmade crafts. Whatever it is, the idea of bringing it to life excites you, but the technical barriers seem daunting.

But web development is just the beginning. Python's versatility extends far beyond building websites. From automating mundane tasks like renaming files or sending emails to developing complex machine learning algorithms, Python empowers you to automate virtually anything you can imagine.

The Art of Automation: Python as Your Personal Assistant

What if I told you Python could save you from tedious tasks? That's right – Python can automate things. Need to organize files? Python can do it. Want to send emails automatically? Python's got your back. It's like having a personal assistant who never gets tired!

Python's Superpowers: What Makes It Special?

- **Versatility:** Python can be used for web development, data analysis, artificial intelligence, and more.
- **Community:** With Python, you're never alone. A vast community of fellow learners and experts is always ready to help.
- **Fun:** Yes, coding with Python is enjoyable. It's like playing with digital LEGO bricks.

Ready to Begin Your Python Adventure?

As we embark on this journey together, remember learning Python is like learning a new language. It takes practice, patience, and a sprinkle of creativity. But the rewards? They're limitless. You're not just learning to code; you're learning to bring ideas to life. So, are you ready to unlock the magic of Python and see where it takes you? Let's go!

INTRODUCTION TO WEB DEVELOPMENT WITH FLASK AND DJANGO

Start by understanding web development. Web development is your canvas, and your tools are technology and creativity. It's an art form where you construct and maintain digital structures—websites and web applications—that exist in the virtual ecosystem of the internet. Like an architect designs buildings and a builder brings them to life, web development involves designing and programming your digital masterpieces.

Building Your Own Digital Kingdom: Web Development with Python

Now, let's talk about building websites. With frameworks like Django and Flask, Python makes web development a breeze. Imagine creating your blog, a photo gallery, or even a little online game. Python gives you the bricks and mortar to build your internet castle.

Flask and Django: Your Tools for the Journey

While HTML, CSS, and JavaScript are essential, creating more complex web applications requires more powerful tools. This is where Flask and Django, two Python-based frameworks, come into play.

- **Flask:** Imagine Flask as a nimble speedboat. It's lightweight, easy to maneuver, and perfect for exploring the smaller islands of the web. Flask allows you to build web applications piece by piece, making it ideal for beginners and small—to medium-sized projects.
- **Django:** Now, think of Django as a majestic cruise ship. It's larger, more powerful, and equipped with everything you need for a long voyage. Django follows a "batteries-included" approach, providing built-in tools for almost everything you need in a web application. It's designed for rapid development and handling complex, large-scale projects.

Why Choose Flask or Django for Web Development?

The beauty of using Flask or Django lies in their Python roots. Python's readability and simplicity make these frameworks accessible to beginners and efficient for experienced developers. Whether you're building a small personal project with Flask or a large-scale application with Django, Python makes the web development process smoother and more enjoyable.

WHAT WEB DEVELOPMENT IS AND WHY IT'S COOL

It's the art and science of bringing ideas to life, from concept to reality and ensuring they thrive in the ever-evolving digital ecosystem.

Imagine you're a digital architect and builder. Just as architects design buildings and builders bring them to life, web development involves designing (web design) and building (web programming) your digital structures. It's about creating, building, and maintaining websites and web applications that run in the mystical world of the internet.

A web developer's toolbox is filled with various powerful tools and technologies, each serving a unique purpose in the construction process. HTML, the backbone of the web, provides the structure; CSS adds style and flair, and JavaScript brings interactivity and dynamism to the mix.

But the magic doesn't stop there. Enter content management systems (CMS) – the unsung heroes of web development. These platforms, like WordPress, Joomla!, and Drupal, streamline the development process, empowering developers to build and manage complex websites easily.

In the fast-paced world of web development, innovation is the name of the game. From responsive design to progressive web apps, the possibilities are endless, and the future is boundless.

Surfing the Wave of Web Development: Why It's the Coolest Digital Adventure

Web Development is like being the captain of a ship in the vast ocean of the Internet. Creating, building, and maintaining websites and web applications is an exhilarating journey. Each site is a new island to explore, and each application is a treasure trove of possibilities waiting to be discovered.

Think about it: you can craft a blog that inspires millions, an online store that sells products globally, or a web app that changes how people work or play. The internet is your playground, and your creations can touch lives worldwide.

Your code can build spaces where people meet, share, shop, learn, and even fall in love. From the most straightforward blog to the most complex social network, web development turns the intangible into something tangible.

A Simple Way To Create Web Applications:

Flask is your gateway to the world of web application development, offering a straightforward and intuitive approach to crafting dynamic online experiences.

Its minimalist design and user-friendly approach liberate developers from the complexities of traditional frameworks. Crafting powerful web experiences becomes a breeze, allowing you to focus on creativity rather than infrastructure concerns.

As a lightweight and flexible web framework written in Python, Flask provides developers a simple yet powerful toolkit for building robust web applications.

With Flask, you can dive straight into the heart of web development, leveraging its simplicity and elegance to turn your visions into reality. So, if you're looking for a hassle-free way to create web applications, look no further than Flask - the simple solution for all your web development needs.

MAXIMIZING DEVELOPMENT SPEED AND RESOURCE UTILIZATION WITH DJANGO

Django is designed with one key goal in mind: to facilitate rapid development. It's like having a team of expert builders working alongside you, each taking care of the complex construction parts, allowing you to focus on customizing your dream project.

With Django, the time from concept to completion is dramatically reduced. You spend less time on the tedious aspects of web development and more on bringing your unique ideas to life.

Security: The Fortified Walls of Your Digital Fortress

In the digital world, security is paramount. Django understands this and comes equipped with robust security features right out of the box.

It's akin to building a fortress with walls already fortified. Django helps protect against common security mistakes, such as SQL injection, cross-site scripting, cross-site request forgery, and click-jacking, making your website fast and safe.

Maintainability: Building with the Future in Mind

Websites are living entities that grow and evolve. Django's design is clean and reusable, promoting maintainability. It encourages writing code once and using it anywhere, reducing redundancy and errors. It's like constructing with Lego blocks - each piece can be used to build something new or rearranged to suit changing needs.

Hassle-Free Development: A Developer's Dream

Django takes care of much of the hassle associated with web development. Imagine having a personal assistant who handles all the paperwork, logistics, and groundwork so you can concentrate on the creative aspect of building your website.

This hassle-free approach means you don't have to reinvent the wheel. You can create complex, database-driven websites without getting bogged down in the minutiae of web development.

Moreover, Django invites developers worldwide to contribute as an open-source framework, ensuring the framework continually evolves, improves, and stays ahead of industry trends.

Exceptional Documentation: Your Guidebook to Web Development

One of Django's most significant assets is its documentation. Comprehensive and well-structured, it's like an extensive guidebook that covers everything from the basics to advanced topics. For newcomers and experienced developers alike, Django's documentation is an invaluable resource for learning and reference.

Support Options: Never Walk Alone

Whether you prefer free community support or professional paid-for services, Django has options for everyone. The wealth of tutorials, forums, Q&A sites, and development services ensure that help is always available, whether you're troubleshooting an issue or seeking to enhance your skills.

John Sonmez: A Synopsis of His Journey in Tech

Early Interests and Career Beginnings

John Sonmez's journey in the tech world began with a profound interest in programming and software development. Like many in the field, his fascination with computers and coding grew as he delved deeper into the subject. He started his career as a software developer, working on various projects that honed his technical skills and deepened his understanding of the industry.

From PHP to Python: My Transformational Journey in Web Development with Django and Flask

When I embarked on my journey into web application development, my tool of choice wasn't Python. Instead, it was PHP – a language that, at the time, I regarded as the undisputed king of web development. I was so entrenched in my PHP ways that even the mere suggestion of an alternative seemed almost heretical.

But, oh, how I was mistaken.

Profoundly mistaken.

The turning point came when my friend Patrick introduced me to Django, a Python web framework. It was a revelation. The tasks I labored over for hours in PHP were transformed into minutes' worth of work with Django. The efficiency was astounding.

Not only did my productivity skyrocket, but the quality of my output did as well. My code became faster, more reliable, and far more stable. The difference was like night and day.

My exploration didn't stop with Django. I soon discovered Flask – another Python framework, but with a twist. Flask was smaller, more modular, and infinitely customizable. After years of working with blunt instruments, it was like being handed a set of precision tools.

The shift to Django and Flask was a game-changer. I bid farewell to PHP without a backward glance.

It's true that PHP still powers a significant portion of the web. However, Python, mainly through Django and Flask, is fast becoming the go-to for modern web application development. Its rise isn't just a trend; it's a paradigm shift in how we think about and execute web development.

Establishing a Personal Brand

His focus on personal development and branding within the tech community sets John apart. Early on, he recognized the importance of not just being a great coder but also of being a well-rounded professional. This realization led him to start his blog, "Simple Programmer," where he shared insights about technical skills, soft skills, career development, and productivity.

Contributions to the Developer Community

John's impact is most notably seen in his efforts to mentor and guide upcoming developers. He has authored books, including "Soft Skills: The Software Developer's Life Manual," which provides a comprehensive guide to non-technical skills crucial in a developer's career. He has also created courses and videos, offering advice on various aspects of tech careers.

THE MAGIC OF PYTHON IN AUTOMATING EVERYDAY TASKS

Imagine having a digital wizard at your command, ready to take over those tedious, everyday tasks you dread. Well, enter Python, the magician of the programming world! With Python, you can automate almost anything you can think of – from organizing your cluttered folders to sending out emails or even getting daily weather updates without lifting a finger.

The Python Spellbook for File Organization

Do you know those thousands of photos and documents scattered across your computer? Python can sort them like a librarian with superpowers. With a few lines of Python code, you can:

- Sort files into folders based on type, date, or size.
- Rename a batch of files in an instant.
- Clean up old files that you no longer need.

Imagine a script running, and in moments, your digital mess is transformed into a neatly organized library. That's Python's magic!

Python as Your Postman

Sending emails can be as repetitive as writing lines in a classroom. Python, however, can be your postman. Whether it's sending out newsletter batches, personalized greetings, or even automated responses, Python scripts can handle it all. You write the rules, and Python delivers the mail without you having to hit 'send' every time.

Your Own Weather Oracle

Ever wish you could predict the weather? Python might not be able to control the weather, but it can fetch forecasts for you. With Python, you can:

- Get daily weather updates automatically.
- Set up alerts for specific weather conditions.
- Even collect historical weather data for your area.

Imagine starting your day with a personalized weather report in your inbox, all thanks to your Python script.

Automating Beyond the Basics

These examples are just the tip of the iceberg. Python's extensive libraries and community-contributed modules make it possible to automate a vast array of tasks:

- Automatically backing up your essential files.
- Organizing your music or movie collection.
- Generating reports and graphs from data.

Python: Your Digital Wand

In this world where time is precious, Python acts like a digital wand, making mundane tasks disappear with a flick. It's a tool that saves you time and allows you to focus on what truly matters – your work, your hobbies, or just enjoying a good cup of coffee while Python handles the rest.

STEP-BY-STEP INSTRUCTIONS FOR A BASIC AUTOMATION SCRIPT

Step 1: Gathering Your Wizarding Tools

Before we start, ensure you have Python installed on your computer – it's like having your wizard's staff. You'll also need a folder filled with files you want to rename. It could be a collection of pictures or documents from your last adventure.

Step 2: Entering the Python Lair

Open your favorite code editor, where you'll cast your Python spell. If you're new to this, IDLE (Python's built-in editor) or a simple text editor will do just fine.

Step 3: Writing the Spell (Script)

Now, let's write our renaming spell. Don't worry; I'll guide you through each line of the magic.

Step 4: Cast Your Spell

Run the script. You can save the file with a .py extension and run it in your command line, terminal, or editor if it supports it.

Step 5: Behold the Magic!

Open your folder and witness the magic – all your files now have new names neatly organized like soldiers in a row. No more chaotic pile of file names!

Extra Tips for Your Wizarding Journey

- Interacting with APIs: In future adventures, you can use Python to talk to various APIs, fetching data like weather forecasts or stock prices.
- Web Scraping: Python can also extract data from web pages, which is perfect for gathering information for your quests.
- Reformatting and Organizing Data: With Python, you can transform and tidy up data, making sense of even the most chaotic datasets.
- Consolidating Tasks: Automate multiple tasks, like sending emails or backing up files, all with a single script.
- Reading and Writing Files: Easily read and write data from one file to another, transforming information as needed.

Reminder!

There you have it, a basic yet powerful Python script to rename files – your first step in the world of automation magic. Remember, this is just the beginning. Python's true power lies in its versatility.

A Student's Journey with Python

Meet Alex, a college student majoring in computer science. Alex juggled a hectic schedule with multiple classes, part-time work, and a social life. Keeping track of assignments, deadlines, and project due dates became challenging. To address this, Alex decided to automate homework reminders using programming skills.

Alex's automation process involved:

Email Integration: Alex set up a dedicated email for academic notifications, receiving assignment due date emails from professors, TAs, and the university's platform.

Scripting: Using Python, Alex automated email parsing to extract assignment details like due dates and course names.

Task Management: Alex utilized Trello or Asana, where Python scripts created tasks for each assignment with relevant details and email links.

Notification System: Alex received alerts for new assignments on his phone and computer by integrating the task tool with notifications.

Custom Reminders: Alex programmed reminders at intervals before due dates, ensuring timely notifications one week, three days, and one day prior.

By automating homework reminders, Alex could effectively manage their workload without worrying about missing deadlines. The process saved time, reduced stress, and allowed Alex to focus on academic success and personal interests.

INTRODUCTION TO COMMAND-LINE TOOLS: CLICK AND ARGPARSE

It's like stepping into a secret chamber of the computer, where you have the authority to run programs, manipulate files, and engage with the machine directly.

A command-line interface (CLI) is a text-based way to interact with your computer, allowing you to run programs, manage files, and control the system directly. These are also known as command-line user interfaces or console user interfaces.

Sounds confusing? But fear not, for Click and Argparse are here to guide you through this digital adventure!

Click adds a layer of user-friendliness to your command-line applications, making them as smooth as butter to work with. Meanwhile, Argparse is your trusty companion for parsing command-line arguments and options, ensuring your tools are intelligent and adaptable.

Say goodbye to the monotony of manual work. Command-line tools are your ticket to automating tasks, saving time, and reducing the chances of making mistakes. You can make your computer do the heavy lifting with some scripting magic.

Crafting Your First Command-Line App with Click

1. Ensure Python and Click Are Installed:

- Make sure Python is installed on your computer. You can download it from python.org.
- Install Click using pip. Open your command line or terminal and run:

```python
import click

@click.command()
def greet():
    """Simple greeting function."""
    click.echo("Hello, magical world of Click!")

if __name__ == '__main__':
    greet()
```

2. Create Your Python Script:

- Open a text editor or an Integrated Development Environment (IDE) like PyCharm or Visual Studio Code.
- Create a new file and name it (e.g., magic_cli.py).
- Write the Click script in the file. Here's a simple example:

3. Save and Run the Script:

- Save the file.
- Open a command line or terminal.
- Navigate to the directory where your file is saved.
- Run the script using Python:

```
Copy code

python magic_cli.py
```

4. View the Results:

- After running the script, you should see the output in the terminal. In our example, it should print:

```
Copy code

Hello, magical world of Click!
```

5. Experiment and Learn:

- Try modifying the script to do different things. Experimenting is a great way to learn more about Python and Click.
- You can add more commands, options, and even user prompts.

Crafting Your Command-Line Adventure with Argparse

If you thought building command-line tools was only for tech gurus, you're in for a pleasant surprise! Argparse makes it a breeze, and I'm here to guide you through it step by step. Let's dive in and start this friendly and fun coding adventure.

Step 1: Set Up Your Script

Create a new Python script, let's call it "greet.py," and open it in your favorite text editor. This will be the home for our command-line application.

Step 2: Import the Argparse Module

At the top of your script, import the Argparse module. This will give you access to all the magic Argparse has to offer.

```python
import argparse
```

Step 3: Define Your Command-Line Arguments

Now, let's define the arguments our script will accept. In this case, we'll create an argument for the user's name.

```python
import argparse

# Step 3.1: Create the parser
parser = argparse.ArgumentParser(description='A friendly greeting progr

# Step 3.2: Define the argument
parser.add_argument('name', type=str, help='Your name')

# Step 3.3: Parse the argument
args = parser.parse_args()

# Step 3.4: Implement your command
print(f"Hello, {args.name}! Welcome to the world of Argparse!")
```

Step 4: Display the Greeting

Finally, let's use the provided name to greet the user.

```python
print(f'Hello, {args.name}! Welcome to the world of command-line applic
```

INTERACTIVE ELEMENT

Task: Create Your Web Page

Dive into the world of web development by creating a basic web page about yourself or your favorite hobby. This fun exercise is your first step into the exciting realm of HTML and CSS, where you'll craft a digital space that reflects your personality or passion.

Step 1: Plan Your Project

Decide on the purpose and functionality of your Python project. Outline what you want it to accomplish.

Step 2: Set Up Your Environment

Install Python and any necessary libraries or frameworks for your project. Set up a virtual environment for better dependency management.

Step 3: Write Your Code

Start coding your project using your preferred editor or IDE. Break down tasks into smaller functions or modules for easier development.

Step 4: Test Your Code

Regularly test your code to catch errors early. Use unit tests and test cases to ensure each component works as expected.

Step 5: Debug and Refine

Debug any errors or issues that arise during testing. Refactor your code for better readability and performance.

Step 6: Document Your Code

Document your code using comments and docstrings to explain its functionality and usage. This will make it easier for others to understand and contribute to your project.

Step 7: Share Your Project

Once your project is complete, please share it with others on platforms like GitHub or PyPI. Encourage feedback and contributions from the community.

SEGUE:

In this chapter, we delved into the fascinating world of web development and automation using Python. From building basic web pages to automating everyday tasks, you've learned how Python can empower you to create valuable tools and applications.

Key takeaways include:

- Python offers a wide range of libraries and frameworks for web development, including Flask and Django.
- Automation with Python can streamline repetitive tasks, saving you time and effort.
- Integrating Python with other tools and platforms enhances productivity and efficiency.

Now, it's time to put your newfound knowledge into action. Consider creating your mini web project or automating a task that could benefit from Python's capabilities.

Looking ahead, we're diving into the exciting realm of Machine Learning and Artificial Intelligence. Imagine the possibilities of teaching your computer to recognize patterns, make decisions, and learn from experiences. Get ready to unleash your creativity and explore the endless opportunities that Python has to offer!

EXPLORING MACHINE LEARNING AND AI

Have you ever wondered how Netflix knows exactly what movies and TV shows you'll love to watch next? Or how do virtual assistants like Siri or Alexa understand you better daily? It's all thanks to Machine Learning and Artificial Intelligence.

Believe it or not, these seemingly magical feats are not the work of sorcery but rather the remarkable capabilities of Machine Learning and Artificial Intelligence (AI). It's a fascinating blend of science, technology, and data-driven insights that powers the intelligent features we've come to rely on in our daily lives.

Machine Learning algorithms continuously learn and improve over time, refining their predictions and responses based on feedback and new data. So, the more you interact with these systems, the better they understand your preferences and anticipate your needs.

BASICS OF MACHINE LEARNING WITH SCIKIT-LEARN

Machine Learning algorithms are like treasure maps that guide computers in their learning journey. They start with historical data – the stories of the past – and use it to make predictions, sort information into categories, group similar things together, and even find the simplest way to describe complex data.

Imagine if your computer could learn from its experiences like you do. That's precisely what Machine Learning (ML) is all about! It's like teaching your computer to become more intelligent by analyzing data loads over time.

Bringing Ideas to Life with ML

And here's where it gets super cool – Machine Learning isn't just about making predictions. It can also help organize data, create new content, and even generate ideas for creative projects. Just look at some of the latest ML-powered apps, like ChatGPT, Dall-E 2, and GitHub Copilot – they're changing the game with their ability to think and create independently!"

A Simple, Relatable Explanation

Suppose you're teaching a friend how to recognize different types of fruit. You show them apples, bananas, oranges each time you name them. Gradually, your friend learns to identify each fruit on their own. This is a lot like machine learning (ML), but instead of fruit, it teaches a computer to recognize patterns and make decisions based on data.

How It Works:

After seeing enough examples, the computer learns to make decisions based on new data it has never seen before. For instance, after looking at many photos of cats and dogs, it can tell which is which in new images.

The more data the computer sees, the better it gets at making decisions. It's like your friend becoming better at recognizing fruits the more you show them. This is why it's called 'machine learning' – the computer is learning and improving over time.

SCIKIT-LEARN: YOUR GATEWAY TO MACHINE LEARNING IN PYTHON

Scikit-learn is like a Swiss Army knife for machine learning in Python, offering an extensive suite of tools for supervised and unsupervised learning.

It isn't just a library; it's a bridge connecting the Python programming language with the power of machine learning. Whether conducting scientific research, working on a cutting-edge commercial project, or just exploring data, scikit-learn offers a comprehensive, accessible platform for all your machine learning needs.

A License to Learn and Innovate

Scikit-learn is open to all under a permissive simplified BSD license. This welcoming approach means it's freely available and widely distributed across numerous Linux distributions. It's a green light for academic exploration and commercial ventures, making it a popular choice in diverse fields.

Built on a Foundation of Scientific Python

At its core, scikit-learn stands on the shoulders of the SciPy (Scientific Python) stack, a collection of libraries for scientific computing in Python. Before diving into scikit-learn, you'll need this stack, which includes:

- NumPy: Consider it the foundational building blocks, supporting multi-dimensional arrays and matrices.
- SciPy Library: This is the engine room for scientific and technical computing.
- Matplotlib: It's your canvas for creating various 2D and 3D visualizations.
- IPython: An enhanced interactive console that supports testing, debugging, and quick iteration.
- Sympy: Dive into symbolic mathematics with ease.
- Pandas: Your go-to for data manipulation and analysis, making data more readable and alterable.

The Birth of Scikit-learn from SciPy

Scikit-learn is one of the many extensions or modules stemming from SciPy, conventionally known as SciKits. It specifically focuses on providing learning algorithms, hence the name scikit-learn.

Expanding Your Machine Learning Horizons

With scikit-learn, you unlock the door to an array of machine-learning possibilities. From building classification models to clustering data, scikit-learn equips you with the tools to analyze and interpret complex datasets, make predictions, and uncover insights that can influence real-world decisions.

Brian Broll: Innovator in Computer Science Education and Creator of NetsBlox

Once upon a time, in the world of technology and education, there was a visionary named Brian Broll. Brian's journey wasn't just a path of personal ambition but one driven by a passion to transform how young minds learn about the complexities of computer science.

Brian's story began in the hallowed halls of academia, where he delved deep into computer science. With its abstract concepts and technical jargon, the world of networked computing often seemed like an insurmountable mountain to young learners. Brian saw not just a challenge but an opportunity.

Driven by a passion for education and a deep understanding of computer science, Brian envisioned a tool to bridge this gap. He imagined a platform that was engaging and accessible, one that could turn complex ideas into interactive, educational adventures. This vision was the seed that eventually grew into NetsBlox.

Brian Broll's involvement with NetsBlox showcases how innovative tools and platforms can be developed to enhance the learning experience in STEM education.

The journey wasn't easy. Brian worked tirelessly, often venturing into uncharted territories, combining the principles of distributed systems with the simplicity of block-based programming.

As NetsBlox came to life, it was as if Brian had given students a magic key to unlock the mysteries of networked computing. The platform allowed them to send messages worldwide, call procedures in distant lands, and collaborate in ways they never thought possible. What once seemed like a distant, complex world was now at their fingertips, waiting to be explored and understood.

Brian's journey reminds us that within the world of technology lies not just circuits and codes but a canvas for creativity, learning, and endless possibilities.

INTRODUCTION TO NEURAL NETWORKS WITH TENSORFLOW AND KERAS

You are creating a brain, but one made of code – that's what neural networks are all about! They're like intricate webs of neurons (but digital) designed to mimic the way our human brains think and learn. With neural networks, you're not just programming; you're teaching your computer to recognize patterns, make decisions, and even solve problems like a detective.

Imagine you're building a super-smart robot. To make it intelligent, you need to give it a brain. Computers create an extraordinary brain called a neural network. It's like making a mini-brain for computers so they can think and learn like we do!

How It Works: Teamwork Makes the Dream Work

Think of each neuron in a neural network as a team player in a big game. They pass information to each other, make decisions, and figure out the best way to solve a problem, just like how you and your friends might work together to solve a mystery.

TensorFlow: Your Magical Toolkit

Enter TensorFlow, Google's open-source library that's a favorite among AI enthusiasts and professionals alike. Think of TensorFlow as your magical toolkit for building and training neural networks. Whether working on a simple project or a complex AI model, TensorFlow brings the necessary muscle and flexibility.

Keras: The Friendly Guide in TensorFlow

Now, meet Keras – your friendly guide in the TensorFlow universe. Keras is an API (Application Programming Interface) that runs on top of TensorFlow. It's like having a wise companion who makes the journey smoother and more intuitive. Keras simplifies the process of building and training neural networks, making them accessible even if you're starting them.

The Excitement Awaits

As you step into this world, remember you're at the forefront of a technological revolution. The road might be steep, but the view from the top is breathtaking. Neural networks are your canvas, and TensorFlow and Keras are your brushes and paints. Let's create something unique together!

BUILDING A BASIC NEURAL NETWORK WITH TENSORFLOW

Let's walk through a fun and straightforward TensorFlow example where we'll build a basic neural network. This network will learn to predict the relationship between numbers. Think of it like teaching your computer to understand a pattern or solve a simple puzzle.

Step 1: Setting the Stage

First, ensure you have TensorFlow installed. If not, you can install it via pip:

```bash
pip install tensorflow
```

Step 2: Write Your Python Script

Now, let's create a Python script. Open your favorite text editor, create a new file, and name it simple_neural_net.py.

Step 3: Import TensorFlow

At the top of your script, import TensorFlow. We'll also import NumPy for handling numerical operations:

```python
import tensorflow as tf
import numpy as np
```

Step 4: Prepare the Data

Let's create some simple data. For instance, the relationship could be "double the input number."

```python
# Input - X values
X = np.array([1, 2, 3, 4, 5, 6], dtype=float)

# Output - Y values
Y = np.array([2, 4, 6, 8, 10, 12], dtype=float)
```

Step 5: Build the Model

Now, let's build a simple neural network with one layer and one neuron.

Step 6: Compile the Model

Before training, we need to compile the model. We'll specify the optimizer and loss function.

```python
model.compile(optimizer='sgd', loss='mean_squared_error')
```

Step 7: Train the Model

Let's train (fit) our model on the data for a specified number of epochs.

```python
model.fit(X, Y, epochs=500)
```

Step 8: Make Predictions

After training, let's use our model to predict the output for a new input.

```python
print(model.predict([10.0]))
```

This line predicts the output if we input ten into our model.

Step 9: Run Your Script

Save your simple_neural_net.py file and run it in your command line or terminal:

```bash
python simple_neural_net.py
```

You've just built and trained your first neural network using TensorFlow. Experiment with different data types, layers, and neurons to see how your predictions change.

BUILDING A BASIC NEURAL NETWORK WITH KERAS

Let's dive into a beginner-friendly example using Keras, a user-friendly interface for building neural networks. We'll create a simple model that learns to predict a basic pattern. Imagine you're teaching your computer a magic trick with numbers!

Step 1: Set the Stage with Keras

First, ensure you have TensorFlow installed, as Keras is part of TensorFlow. If not installed, you can do so via pip:

```bash
pip install tensorflow
```

Step 2: Create Your Python Script

Open your favorite text editor, create a new file, and name it simple_keras_model.py.

Step 3: Import TensorFlow and Keras

Start your script by importing TensorFlow. Keras is part of TensorFlow, so you don't need a separate import for Keras:

```python
import tensorflow as tf
```

Step 4: Prepare the Data

For this example, we'll use a simple dataset. Let's say the relationship we want to learn is "triple the input number."

```python
# Input - X values
X = [1, 2, 3, 4, 5, 6]

# Output - Y values
Y = [3, 6, 9, 12, 15, 18]
```

Step 5: Build the Neural Network Model

Now, let's construct a basic neural network model using Keras. We'll create a model with one layer and one neuron.

```python
# Create a Sequential model with one Dense layer
model = tf.keras.Sequential([tf.keras.layers.Dense(units=1, input_shape
```

Step 6: Compile the Model

We must compile the model by specifying an optimizer and a loss function.

```python
model.compile(optimizer='sgd', loss='mean_squared_error')
```

Step 7: Train (Fit) the Model

It's time to train our model on the dataset we created.

```python
model.fit(X, Y, epochs=500)
```

Step 8: Make Predictions

After training, let's use our model to predict the output for a new input.

```python
print(model.predict([7]))
```

This line predicts the output when the input is 7.

Step 9: Run Your Script

Save your simple_keras_model.py file and run it in your command line or terminal:

```bash
python simple_keras_model.py
```

You've just created and trained your first neural network using Keras.

GETTING STARTED WITH PYTORCH

PyTorch is a rising star in the universe of machine learning! PyTorch is not just a tool; it's a gateway to turning the complex ideas of machine learning into reality. Let's embark on an adventure to explore PyTorch and its role in AI and deep learning.

PyTorch: Your Torch in the Machine Learning Darkness

PyTorch is an open-source machine learning framework that has become the torchbearer for many in AI research and development. Built upon the foundations of the Python programming language and the Torch library, PyTorch is like a powerful spellbook for wizards delving into machine learning and intense neural networks.

The Origin: From Torch to PyTorch

The story of PyTorch begins with Torch, an earlier ML library cherished for its flexibility and speed but written in the Lua scripting language. PyTorch takes the essence of Torch and blends it with the simplicity and popularity of Python, making it more accessible and powerful.

Why PyTorch Shines Brightly

PyTorch has rapidly become a beloved platform among deep learning researchers for several reasons:

- Research and Deployment Harmony bridges the gap between research prototyping and production deployment. PyTorch is like a fast train that smoothly transitions your experimental ideas into real-world applications.
- Dynamic Computational Graphs: Unlike some other frameworks, PyTorch uses dynamic computational graphs, making it more intuitive and flexible. It's like molding clay in your hands – you can change the shape of your models as you go.
- Pythonic Nature: PyTorch feels more like writing Python code, making it a natural fit for Python enthusiasts and a friendlier learning curve for beginners.
- Community and Ecosystem: With a thriving community, PyTorch not only offers extensive libraries and tools but also a wealth of knowledge, tutorials, and support.

The PyTorch Adventure: What Lies Ahead

As you begin your journey with PyTorch, you'll discover how to breathe life into arrays and tensors, train neural networks to learn from data, and even see them make predictions. Whether you're interested in image recognition, language processing, or creating AI that can play games, PyTorch is your trusted companion on this journey.

FUN PYTORCH PROJECT: IMAGE CLASSIFICATION WIZARDRY WITH TRANSFER LEARNING

Ready to embark on an exciting PyTorch project? Let's dive into the world of image classification using transfer learning. It's like giving your computer a pair of magical glasses to see and understand pictures!

The Magic of Transfer Learning

Transfer learning is like teaching your computer a new trick using the knowledge it already has. It's about taking a model that's already learned a lot about one thing (like recognizing cats and dogs) and teaching it to apply that knowledge to something new (like identifying different documents). It's efficient and pretty cool!

Your Mission: Classify Images with PyTorch

Imagine you have many images – driving licenses, social security cards, and more. Your task? Build a model that can look at these images and tell what kind of document each one is. We'll use PyTorch and a super-smart model called ResNet to do this.

Step 1: Meet ResNet, Your AI Assistant

ResNet is a powerful model that's already learned a lot about images. It's like a wise old professor of pictures. We'll take this model and add our twist to make it great at recognizing different documents.

Step 2: Gather Your Images

Our magical ingredients are images of driving licenses and other documents. These images are of different shapes and sizes—just like in the real world.

Step 3: Prep Your Potions (Preprocess the Images)

Before we start training, we need to prepare our images. This means ensuring they're all the same size and format—it's like chopping up ingredients before cooking.

Step 4: Train Your Model with PyTorch

Now comes the fun part! We'll take our prepped images and feed them to our model. Using PyTorch, we'll teach our model to understand what each image is. It's a bit like teaching a parrot new words.

Step 5: Watch the Magic Happen

After training, you can show your model a new image it's never seen, and it'll use its newfound knowledge to tell you what it is. Is it a driving license or a social security card? Your model will now be able to guess!

PYTORCH PROJECT FOR BEGINNERS: CRAFTING A LOGISTIC REGRESSION MODEL

Hey, young coders, now you'll build a Logistic Regression model using PyTorch from scratch! Picture yourself as a digital artisan, crafting a model that's not just about crunching numbers but about understanding probabilities and making predictions. Our mission? To dive into the world of binary image classification. Sounds like a quest from a sci-fi novel, right?

What is Logistic Regression?

Logistic Regression is like a fortune teller in the world of machine learning. It doesn't just predict outcomes; it calculates the probability of something happening based on given information. For example, it can indicate if a picture is of a cat or not a cat. It's perfect for yes/no (binary) questions.

Your Mission: Binary Image Classification

Your quest is to build a model that looks at images and answers a simple question: is this one thing or another? We'll use PyTorch to create a model that understands pictures and makes predictions about them.

Step 1: Clean and Prep Your Data

Before our model can learn, we need to tidy up our data, like cleaning your room before starting a new project. We'll make sure our images are the right size and format. This step is crucial for helping our model learn efficiently.

Step 2: Building the Logistic Regression Model

Now, we put on our builder's hats and start constructing our model. We'll use PyTorch to lay the foundations and build our logistic regression model. It's like assembling a puzzle—each piece needs to fit perfectly.

Step 3: Pretesting the Model

Before we go full steam ahead, let's do a pretest, like a dress rehearsal. We'll run initial tests to see if our model looks at the data correctly. It's a bit like tasting your food while cooking.

Step 4: Train Your Model

Here's where the real magic happens! We'll train our model with prepared images, teaching it to understand and predict. Training a model in machine learning is like training for a sport – the more practice you get, the better you perform.

SETTING SAIL ON HYPERPARAMETER TUNING ODYSSEY WITH PYTORCH

Imagine you're a pilot fine-tuning the controls of a spaceship for the smoothest, fastest journey through the stars. That's what hyperparameter tuning does for neural networks in machine learning.

Hyperparameters: The Secret Knobs of Neural Networks

In the universe of neural networks, hyperparameters are like secret knobs and dials. They control how the network learns but do not learn from the data. Instead, you, the brilliant coder, set them. We're talking about how many epochs to train for, the learning rate when to stop training early to avoid overfitting (early stopping), and dropout rates to prevent the network from relying too much on specific neurons.

Your Mission: Optimize for Stellar Performance

Using PyTorch, we'll explore how to understand and optimize these settings. It's like fine-tuning a race car's engine and suspension for the best possible performance on the track.

Step 1: Understand Your Controls

First, you'll dive into what each hyperparameter does.

- Epochs: How many times the network sees the entire dataset.
- Learning Rate: How fast it learns; too slow can take forever, and too fast can overshoot the goal.
- Early Stopping: To stop training before it starts getting worse.
- Dropout: Randomly ignores neurons during training to improve robustness.

Step 2: Begin the Tuning Adventure with PyTorch

With PyTorch, you'll start experimenting:

- Adjust the learning rate and observe the changes.
- Play with the number of epochs and see how it affects learning.
- Implement early stopping and dropout to tackle overfitting.

Step 3: Watch the Magic Happen

As you tweak these hyperparameters, watch your neural network evolve. It's like training a dragon; with the proper care and training, it becomes more powerful and intelligent.

INTERACTIVE ELEMENT

Let's see how much you've absorbed from the chapter!

AI Quiz: Test Your Machine Learning Knowledge

1. What is a neural network in the context of machine learning?

A) A computer network
B) A brain-like system in computers
C) A programming language
D) A type of computer virus

2. What does 'training a model' in machine learning mean?

A) Fixing bugs in the model
B) Programming the model to perform tasks
C) Feeding data to the model so it can learn
D) Connecting the model to the internet

3. What is 'overfitting' in machine learning?

A) When a model performs too well on training data
B) When a model doesn't fit into the computer's memory
C) When the model's size is too large
D) When the model overheats during training

4. What role does 'dropout' play in a neural network?

A) It deletes unnecessary data
B) It randomly ignores some neurons during training
C) It drops the learning rate
D) It stops the network from working

5. In machine learning, what is a 'hyperparameter'?

A) a very important parameter
B) A high-level programming parameter
C) A value set before training a model
D) A parameter used for hyper-speed computing

Answers

1: B

2: C

3: A

4: B

5: C

SEGUE

As we close this chapter, let's take a moment to marvel at our embarked journey. You've unlocked the mysteries of Neural Networks, TensorFlow, Keras, and PyTorch – tools at the forefront of AI and machine learning. From understanding the basics of neural networks to tuning hyperparameters for optimal performance, you've opened the door to a world where computers learn and think.

Key takeaways from this chapter include:

- Python's ease of use and readability make it ideal for beginners and AI and machine learning experts.
- Python offers a rich ecosystem of libraries and frameworks like TensorFlow, scikit-learn, and PyTorch, which empower developers to create powerful AI applications.
- We discussed the significance of data in AI and how Python helps in data collection, preprocessing, and analysis, which are crucial steps in building AI models.
- You learned about the importance of model training and how Python's libraries provide tools for model selection, optimization, and evaluation.

- Finally, we touched upon the potential of Python in AI ethics and the responsible development of AI systems.

Start by experimenting with Python, exploring the discussed libraries, and applying your knowledge to real-world projects to enhance your AI and machine learning skills.

In the next chapter, we'll explore the world of Python's advanced applications. Imagine using Python to automate everyday tasks, decode the complexities of human language, or even create video games.

ADVANCED PYTHON APPLICATIONS

D id you know that Python can do some seriously cool stuff? It can help keep computers safe from hackers, teach tiny gadgets to do amazing things, and even make computers understand human language! It sounds like something out of a sci-fi movie, right?

Envision a scenario where Python, the language of choice for developers, is a formidable shield against cyber intruders, unlocks boundless potential in miniature devices, and facilitates seamless communication between man and machine.

Now that we've wrapped up our journey through neural networks and machine learning, it's time to explore Python's advanced applications. Think of Python as a versatile tool that can do almost anything you can dream up!

SCRIPTING FOR CYBERSECURITY AND ETHICAL HACKING

Think of cybersecurity as a superhero for computers. Just like a city needs protection from villains, our computers, smartphones, and networks need protection from digital bad guys—hackers. These hackers are like sneaky burglars trying to steal information or vandalize your digital home (computer and online accounts).

Ethical Hacking: The Good Wizards of the Cyber World

Ethical hacking might sound like a contradiction – how can hacking be good? Ethical hackers are like secret agents hired to find weaknesses in a computer system, but instead of causing harm, they help fix these problems. They are the excellent wizards who use their powers to find and fix the cracks in our digital walls.

Simplified Concepts

- **Firewalls**: Like a castle's walls protecting it from invaders, firewalls protect your computer from unauthorized access from the internet.
- **Antivirus Software:** This is like having a doctor who checks for viruses and cures them before they can make your computer sick.
- **Phishing**: Imagine fishermen using fake, tasty bait to catch fish. In the digital world, phishing is when hackers use counterfeit emails or messages as bait to trick you into giving them your personal information.
- **Encryption:** This is like writing a secret letter in a code that only you and the person you're sending it to can understand. It keeps your data safe even if someone else finds it.

By viewing cybersecurity and ethical hacking through these simple yet engaging lenses, we can better understand their importance in our increasingly connected world. It's about building digital shields and having skilled allies to protect our precious online information from the cyber villains lurking in the shadows.

Let's look at some basic coding examples that illustrate cybersecurity and ethical hacking concepts in an engaging and simplified manner.

1. Creating a Simple Password Checker in Python

This script will help you understand how important strong passwords are for cybersecurity.

```python
def check_password_strength(password):
    if len(password) < 8:
        return "Weak: Password too short!"
    elif not any(char.isdigit() for char in password):
        return "Weak: Password needs a number!"
    elif not any(char.isupper() for char in password):
        return "Weak: Password needs an uppercase letter!"
    else:
        return "Strong Password!"

user_password = input("Enter your password to check its strength: ")
print(check_password_strength(user_password))
```

2. Basic Port Scanner with Python

Ethical hackers often use port scanners to check for open ports in a system, which can be entry points for hackers. Here's an elementary example:

```python
import socket

def scan_port(ip, port):
    scanner = socket.socket(socket.AF_INET, socket.SOCK_STREAM)
    scanner.settimeout(1)
    try:
        scanner.connect((ip, port))
        return True
    except:
        return False

ip_to_test = "192.168.1.1"  # Replace with the IP you want to check
for port in range(1, 1025):
    if scan_port(ip_to_test, port):
        print(f"Port {port} is open on {ip_to_test}!")
```

3. Simple Encryption and Decryption in Python

This example shows how data can be encrypted (made secret) and then decrypted (made readable again):

```python
from cryptography.fernet import Fernet

# Generating a key and instancing a Fernet object
key = Fernet.generate_key()
cipher_suite = Fernet(key)

# Encrypting a message
def encrypt_message(message):
    return cipher_suite.encrypt(message.encode())

# Decrypting a message
def decrypt_message(encrypted_message):
    return cipher_suite.decrypt(encrypted_message).decode()
```

```
original_message = "This is a secret message!"
encrypted = encrypt_message(original_message)
decrypted = decrypt_message(encrypted)

print(f"Original Message: {original_message}")
print(f"Encrypted Message: {encrypted}")
print(f"Decrypted Message: {decrypted}")
```

These examples provide a fundamental yet practical insight into some of the coding aspects of cybersecurity and ethical hacking. While simplified, they demonstrate the principles of securing data, scanning for vulnerabilities, and the importance of encryption in protecting information.

BASIC SCRIPTING FOR CYBERSECURITY USING PYTHON

Imagine Python as a magical wand. With its simple, readable syntax – like an incantation's clear, concise words – Python allows you to perform complex cybersecurity spells efficiently. Python scripting is essential in any cyber mage's arsenal, from banishing malicious software to conjuring protective barriers around sensitive data.

Scripting Spells for Security

- **Enchanting Network Scanners:** With Python, you can create spells (scripts) that scan through the vast digital landscapes, detecting potential intruders and vulnerabilities. It's like sending out scouts into the night, ever-vigilant for signs of enemy forces.
- **Potion of Data Analysis:** Python helps you brew powerful potions (scripts) to analyze mountains of data. These potions can sift through endless digital scrolls to find

hidden patterns and clues, revealing the secrets of cyber threats.

- **Shielding Against Dark Arts:** Python enables you to craft powerful shields (firewalls and intrusion detection systems) that block hackers' dark magic. These shields stand guard at the gates of your digital kingdom, repelling the spells of the cyber underworld.
- **Automated Incantations:** Automate repetitive tasks with Python's scripting magic. Like summoning a legion of tireless golems, these scripts tirelessly monitor, log, and respond to potential threats, ensuring your defenses never sleep.

Your Grimoire (Scripting Guide)

- **Begin with the Basics**: Start your journey by learning Python's syntax – the words of your spells. Understand variables, loops, and functions, the fundamental incantations of your scripting grimoire.
- **Craft Your First Spell:** Create a simple script, perhaps a potion that alerts you when someone tries to enter your digital domain unauthorized. This "Hello World" of cybersecurity scripting is your first step into a larger world.
- **Advance to Complex Enchantments:** As your knowledge deepens, so does your magical repertoire. Move on to more complex scripts that analyze network traffic, detect anomalies, or even automate responses to cyber threats.

The Journey Ahead

The path of a Python scripting mage is both exciting and challenging. As you delve deeper into the arcane arts of cybersecurity, remember that each line of code is a step towards mastering this digital sorcery. Your journey will be filled with learning, discovery, and the satisfaction of using your powers to protect the digital realm.

So, arm yourself with Python, a young wizard, and prepare to cast spells to fortify the cyber world. The adventure awaits, and the future of digital security rests with skilled sorcerers like you.

Python in IoT with MicroPython

The Internet of Things, often abbreviated as IoT, is like a digital spiderweb connecting various devices to the cloud. These devices, which range from smart appliances to industrial machinery, are equipped with sensors and software to communicate seamlessly, exchange data, and enable various functions and applications.

Picture a world where your toaster chats with your coffee maker, your fridge shares shopping lists with your smartphone, and your thermostat adjusts itself based on your schedule.

This is the Internet of Things (IoT) in action—a vibrant network of intelligent and straightforward devices, all connected and exchanging data with each other and the cloud. From smart sensors to sophisticated software, IoT devices blend seamlessly into our daily lives, transforming ordinary objects into intelligent, interconnected marvels.

MicroPython is a miniaturized, streamlined version of Python specially designed to fit into tiny devices. It's easy to learn and use, but it's compact enough to run on small gadgets like microcontrollers—the brains of IoT devices.

- **Lights, Camera, Automation!:** Think of an intelligent light bulb. With MicroPython, you can program it to change colors or turn on/off at specific times. It's like having a little wizard inside each bulb, listening to your commands, and creating the perfect ambiance.
- **Your Health Assistant:** Your fitness tracker, powered by a microcontroller running MicroPython, counts your steps, monitors your heart rate, and even nudges you to move around if you've been sitting too long. It's like a tiny coach on your wrist!
- **Smart Gardening:** Imagine an intelligent irrigation system in your garden. MicroPython can help program the system to water your plants precisely when needed based on soil moisture data. Your garden will now be self-caring, almost as if it had a green thumb.
- **Home at Your Fingertips:** With MicroPython, your home appliances can communicate with each other. Your alarm clock can tell your coffee maker to start brewing coffee right as you wake up. It's like having a team of tiny robots making sure your morning starts smoothly.

In this fun, interconnected world of IoT, MicroPython is the magic spell that brings inanimate objects to life, making them intelligent and responsive. It's a small step into a larger, more automated world where every device works together to make our lives easier, more efficient, and fun!

MICROPYTHON: SIMPLIFYING HARDWARE CONTROL AND BEYOND

MicroPython isn't your average programming language – it's a tiny powerhouse that runs on small embedded development boards, turning them into dynamic tools for controlling hardware. No more wrestling with complex low-level languages like C or C++; with MicroPython, you can quickly write clean, simple Python code that commands your devices.

But here's the kicker: MicroPython isn't just for seasoned programmers. Its user-friendly interface and Python's intuitive syntax make it a dream come true for beginners dipping their toes into the world of programming and hardware. MicroPython's familiarity and versatility will keep you hooked and eager to explore, even if you're a Python pro.

What sets MicroPython apart from the crowd? Let's dive into its unique features:

Interactive REPL: Say goodbye to the hassle of compiling and uploading code. With MicroPython's Interactive REPL, you can connect to a board and execute code on the fly – perfect for quick learning and experimentation with hardware!

Extensive Software Library: Like its big sibling, Python, MicroPython comes fully loaded with a treasure trove of built-in libraries. From parsing JSON data to network socket programming, MicroPython has you covered with tools for every task imaginable.

Extensibility: MicroPython offers the best of both worlds for those craving even more power. Combining its extensibility with C/C++ allows you to integrate expressive high-level code with lightning-fast low-level functions seamlessly.

So whether you're a beginner itching to bring your projects to life or a seasoned veteran seeking new challenges, MicroPython is your ticket to unlocking a world of endless possibilities in embedded systems.

Transforming Everyday Objects into Smart Devices

Here are some realistic examples of everyday objects that you can control and interact with using MicroPython:

1. LED Light Control:

- *Everyday Object:* LED Bulb
- *Coding Input:*

```python
from machine import Pin
import time

led = Pin(2, Pin.OUT)  # Pin 2 is connected to the LED

while True:
    led.value(1)  # Turn LED on
    time.sleep(2)  # Wait for 2 seconds
    led.value(0)  # Turn LED off
    time.sleep(2)  # Wait for 2 seconds
```

Coding Output: The LED bulb connected to pin 2 turns on and off, alternatively every 2 seconds.

2. Smart Doorbell:

- *Everyday Object: Push Button*
- *Coding Input:*

```python
from machine import Pin
import time

button = Pin(0, Pin.IN)  # Pin 0 is connected to the push button

while True:
    if button.value() == 0:
        print("Someone is at the door!")
        time.sleep(0.5)  # Debounce delay
```

Coding Output: Whenever the push button connected to pin 0 is pressed, the message "Someone is at the door!" is printed. A debounce delay of 0.5 seconds is added to prevent multiple detections from a single press.

Let's explore a simple project idea: creating a Smart Room Temperature Monitor using MicroPython. This project involves using a temperature sensor to monitor the temperature of a room and display the current temperature on a small screen. If the temperature exceeds a set threshold, it will trigger an alert.

PROJECT OVERVIEW: SMART ROOM TEMPERATURE MONITOR

Objective

To create a device that constantly monitors room temperature and provides real-time updates. It will alert the user if the temperature exceeds or falls below certain preset limits.

Components Needed

- A microcontroller compatible with MicroPython (like an ESP32 or ESP8266)
- A temperature sensor (like the DS18B20 or DHT11/DHT22)
- An OLED display (optional for displaying the temperature)
- Breadboard and jumper wires
- Buzzer or LED (for alert)

Steps to Build

Setting Up MicroPython on the Microcontroller:

- Flash MicroPython firmware onto the microcontroller.
- Establish a serial connection for programming.

Connecting the Temperature Sensor:

- Connect the temperature sensor to the microcontroller using jumper wires. Make sure to connect the data, power, and ground pins correctly.

Programming the Microcontroller:

- Write a MicroPython script to read temperature data from the sensor.
- Display the current temperature on the OLED screen.
- Set a threshold temperature. If the room temperature exceeds this threshold, trigger the buzzer or LED.

Testing and Calibration:

- Test the device in different temperature settings.
- Calibrate the sensor if necessary for more accurate readings.

Sample Code Snippet

Here's a simplified version of what the MicroPython code might look like:

```python
import dht
import machine

# Set up the sensor
sensor = dht.DHT11(machine.Pin(4))  # The number 4 is where you connect

def check_temperature():
    sensor.measure()
    temp = sensor.temperature()
    print("It's", temp, "degrees Celsius right now!")

# Let's check the temperature!
check_temperature()
```

Seeing the Temperature:

- When you run this code, your microcontroller will ask the sensor what the temperature is and show you the result.

Play and Learn:

- Try changing where you put the sensor – it may be near a window or in the fridge. See how the temperature changes!

This project is an excellent start for beginners. It's like making your little weather station for your room! Remember, it's okay to make mistakes and ask for help. That's how you learn and get better. Have fun with your temperature checker!

Project Expansion

- Connect the device to Wi-Fi and send temperature alerts via email or a mobile app.
- Log temperature data over time to track climate trends in the room.
- Add humidity monitoring for a more comprehensive environmental check.

NATURAL LANGUAGE PROCESSING WITH NLTK AND SPACY

Natural Language Processing (NLP) gives computers the super-power to understand human speech and text—just like magic! It's a crucial part of artificial intelligence (AI) that has existed for over 50 years, with its roots deeply embedded in the fascinating world of linguistics.

From decoding medical jargon to powering search engines and revolutionizing business insights, NLP works its enchantment across a broad spectrum of real-world applications.

Natural Language Processing (NLP) is like teaching computers to understand and speak human language. It's a bit like learning a new language but for computers. Two of the most popular tools for this are NLTK (Natural Language Toolkit) and spaCy. They are like the language textbooks for computers, filled with exercises and lessons in linguistics.

Let's demystify NLP with some fun and engaging examples using NLTK and spaCy:

1. The Detective of Sentiments: Sentiment Analysis

Imagine your computer could read a book or a tweet and tell you if it's happy, sad, or angry. That's sentiment analysis. You feed text to the computer, and it tells you the mood of the words.

Using NLTK:

- You might use NLTK to scan through movie reviews. The computer can analyze these reviews and tell you which ones are positive and negative.
- It's like having a little critic inside your computer who reads and rates the movies for you!

```python
from nltk.sentiment import SentimentIntensityAnalyzer

# Initialize the sentiment analyzer
sia = SentimentIntensityAnalyzer()

text = "I love this movie, it was fantastic and thrilling!"

# Get sentiment scores
sentiment = sia.polarity_scores(text)

print("Sentiment Scores:", sentiment)
```

2. The Word Magician: Part-of-Speech Tagging

Part-of-speech tagging is where the computer reads a sentence and identifies nouns, verbs, adjectives, etc. It's like teaching a computer to understand the parts of a sentence.

Using spaCy:

- Give spaCy a sentence like "The quick brown fox jumps over the lazy dog," and it will tell you 'quick' and 'brown' are adjectives, 'fox' and 'dog' are nouns, and so on.
- It's like your computer is a grammar teacher, breaking down the sentence structure for you.

```python
import spacy

# Load the English model
nlp = spacy.load("en_core_web_sm")

text = "The quick brown fox jumps over the lazy dog."

# Process the text
doc = nlp(text)

# Print part-of-speech tags
for token in doc:
    print(f"{token.text}: {token.pos_}")
```

3. The Puzzle Solver: Word Tokenization

Tokenization is chopping up text into pieces, called tokens. Think of it like slicing a pizza. Each slice (or token) is a word or a sentence from the text.

Using NLTK:

- Feed an enormous paragraph to NLTK, and it will slice it into individual words or sentences for you.
- This is helpful to analyze or play with specific text parts.

```python
from nltk.tokenize import word_tokenize

text = "NLTK makes text processing so easy and enjoyable!"

# Tokenize the text
tokens = word_tokenize(text)

print("Tokens:", tokens)
```

4. The Language Explorer: Named Entity Recognition (NER)

NER is like a scavenger hunt in text. The computer looks for names of people, places, organizations, dates, and more.

Using spaCy:

- If you give spaCy a news article, it can pick out names of people, places, and even dates.
- Imagine feeding a history book to spaCy, which lists all the important names and places for you – an excellent study tool!

```python
import spacy

# Load the English model
nlp = spacy.load("en_core_web_sm")

text = "Albert Einstein was born in Germany."

# Process the text
doc = nlp(text)

# Extract named entities
for entity in doc.ents:
    print(f"{entity.text}: {entity.label_}")
```

In summary, NLTK and spaCy make it possible to do these amazing things with text. They turn your computer into a language expert, helping you understand, analyze, and predict human language. It's a bit like giving your computer a course in linguistics and watching it become a language whiz!

PROJECT: EXTRACTING NAMED ENTITIES FROM NEWS ARTICLES

Let's walk through an introductory Natural Language Processing (NLP) project using spaCy. We'll create a simple project that extracts named entities (like names of people, places, and organizations) from a text. This is a common NLP task and a great way to get familiar with what spaCy can do.

Objective:

To identify and categorize named entities in news articles. This could be useful for summarizing content, categorizing articles, or extracting useful information quickly.

Steps:

Setup Environment

- Install spaCy: Run pip install spaCy in your command line.
- Download the English model: After installing spaCy, download the English language model by running python -m spacy download en_core_web_sm.

Choose a News Article

- For this project, select a news article of your choice. You can use online news sources or any textual content.

Write Python Script to Process the Text

- You'll write a Python script that uses spaCy to process the chosen news article and extract named entities.

Python Script:

```python
import spacy

# Load the English model
nlp = spacy.load("en_core_web_sm")

# Sample text (replace this with your news article)
text = """Google CEO Sundar Pichai introduced the new Pixel

# Process the text
doc = nlp(text)

# Extract entities
for entity in doc.ents:
    print(f"{entity.text}: {entity.label_}")
```

Running the Script:

- Run the script with your Python interpreter.
- The output will list the entities in the text and their categories (like PERSON for people's names, ORG for organizations, GPE for countries, cities, etc.).

Exploring Further:

- Try using different types of texts to see how the model performs.
- Experiment by extracting other types of information, like noun chunks or verb phrases.
- You can also visualize the named entities using spaCy's displaCy module for a more interactive experience.

Example of Visualization:

```python
from spacy import displacy

# Using displacy to visualize named entities
displacy.serve(doc, style="ent")
```

This project gives you a taste of how powerful NLP tools like spaCy can extract meaningful information from text. It opens doors to more complex applications such as sentiment analysis, text summarization, and language generation. Remember, this is just the beginning, and there's much more you can do with NLP and spaCy!

EMBARKING ON A JOURNEY: THE RISE OF MATHIS ANDRÉ AND FAQBOT

Meet Mathis André, a French teenager whose path to success took an unexpected turn when he dropped out of school at just 16. Initially drawn to website development, André's interest shifted towards bots – software with immense customer service and e-commerce potential.

Fast-forward to today. André, now 17 and based in Brussels, co-founded Faqbot, a pioneering venture focused on revolutionizing traditional 'frequently asked questions' (FAQ) pages through chatbots.

"At Faqbot, we aim to streamline the user experience by trans-forming lengthy FAQs into interactive chatbots," explains André. "It's a challenging task, as accuracy is paramount, and machine learning technology still faces hurdles in this domain."

While the initial hype surrounding chatbots may have waned, the market continues to thrive, with projections indicating substantial growth in the coming years. Despite the evolving landscape, Faqbot remains at the forefront, continually refining its software to adapt to changing user needs.

But the journey hasn't been without its bumps. André's foray into chatbots began after meeting his future Faqbot co-founder, Denny Wong, at a TedX conference. Together, they embarked on various projects, including developing a website builder bot, before channeling their efforts into Faqbot.

Today, Faqbot is a testament to André's determination and innovation, with plans for future expansion already in the pipeline. And while the road ahead may be filled with challenges, André remains undeterred, armed with the invaluable lessons learned along the way.

INTERACTIVE ELEMENT

Cyber Treasure Hunt: A Python Security Puzzle Adventure

You have stumbled upon an old, abandoned computer system rumored to contain valuable cybersecurity secrets. To access the treasure trove of knowledge, you must decipher clues and overcome challenges left behind by the system's enigmatic creator.

Challenge 1: Crack the Firewall

The first challenge is to bypass the system's firewall. You have been provided a Python script (firewall_bypass.py) containing the firewall's algorithm. Analyze the code, find the loophole, and execute the script to turn off the firewall.

```python
# firewall_bypass.py
def bypass_firewall():
    # Firewall algorithm
    print("Firewall bypassed! Access granted.")

# Your task: Analyze and execute the script to bypass the firewall
bypass_firewall()
```

Challenge 2: Decrypt the Access Key

With the firewall disabled, you gain access to the system. However, the treasure is protected by an encrypted access key. You will find a file (access_key.txt) containing the encrypted key. Write a Python script to decrypt the access key and gain entry to the treasure vault.

```python
# access_key.txt
encrypted_key = "hgy XliWevmrk 5×7*"
# Your task: Write a Python script to decrypt the access key
def decrypt_access_key(encrypted_key):
    decrypted_key = ''
    for char in encrypted_key:
        if char.isalpha():
            decrypted_key += chr(ord(char) - 3)
        else:
            decrypted_key += char
    return decrypted_key

access_key = decrypt_access_key(encrypted_key)
print("Decrypted Access Key:", access_key)
```

Challenge 3: Navigate the Maze

Upon entering the treasure vault, you encounter a virtual maze guarded by security measures. Write a Python program (maze_-solver.py) to navigate the maze and reach the treasure chamber. Beware of traps and dead-ends!

```python
# maze_solver.py
maze = [
    ['#', '#', '#', '#', '#', '#', '#', '#', '#', '#'],
    ['#', ' ', ' ', ' ', '#', ' ', ' ', ' ', ' ', '#'],
    ['#', ' ', '#', ' ', '#', ' ', '#', '#', ' ', '#'],
    ['#', ' ', '#', ' ', ' ', ' ', ' ', '#', ' ', '#'],
    ['#', ' ', '#', '#', '#', '#', '#', '#', ' ', '#'],
    ['#', ' ', ' ', ' ', '#', ' ', ' ', ' ', ' ', '#'],
    ['#', '#', '#', ' ', '#', '#', '#', '#', ' ', '#'],
    ['#', ' ', ' ', ' ', ' ', ' ', ' ', ' ', ' ', '#'],
    ['#', '#', '#', '#', '#', '#', '#', '#', '#', '#']
]

def solve_maze(maze):
    # Your maze-solving algorithm here
    pass

# Your task: Write a Python program to solve the maze
solve_maze(maze)
```

Challenge 4: Crack the Vault Code

At the heart of the treasure chamber lies a vault protected by a numeric code. You discover a set of encrypted messages (vault_-codes.txt) that might contain clues to the code. Decrypt the messages and crack the vault code to unveil the ultimate prize.

```python
# vault_codes.txt
encrypted_codes = [
    "jfEgW oeqPkf?",
    "shC sldo?G"
]
# Your task: Write a Python script to decrypt the vault codes
def decrypt_vault_codes(encrypted_codes):
    decrypted_codes = []
    for code in encrypted_codes:
        decrypted_code = ''
        for char in code:
            if char.isalpha():
                decrypted_code += chr(ord(char) - 2)
            else:
                decrypted_code += char
        decrypted_codes.append(decrypted_code)
    return decrypted_codes

decrypted_vault_codes = decrypt_vault_codes(encrypted_codes)
print("Decrypted Vault Codes:", decrypted_vault_codes)
```

Challenge 5: Secure the Treasure

Congratulations! You've reached the treasure, but your journey isn't over yet. Write a Python script to encrypt the treasure's contents (treasure.txt) using advanced encryption techniques and secure it from unauthorized access.

```python
# treasure.txt
treasure_contents = "This is the treasure contents. Keep it safe!"
# Your task: Write a Python script to encrypt the treasure contents
def encrypt_treasure(contents):
    encrypted_contents = ''
    for char in contents:
        encrypted_contents += chr(ord(char) + 3)
    return encrypted_contents

encrypted_treasure = encrypt_treasure(treasure_contents)
print("Encrypted Treasure:", encrypted_treasure)
```

SEGUE

This chapter completed an exciting journey into Python cyberse-curity and interactive puzzles. We explored how Python can deci-pher encrypted messages, identify vulnerabilities in code, and solve real-world cybersecurity challenges. Along the way, we honed our Python scripting skills and gained valuable insights into the principles of digital security.

Key takeaways from this chapter include:

- Python scripting can be a powerful tool for cybersecurity, enabling us to decrypt messages, analyze code, and identify security vulnerabilities.
- By practicing with interactive puzzles, we can sharpen our Python skills while learning about essential concepts in cybersecurity.
- Implementing the ideas in this chapter can help us become more proficient in Python and better equipped to tackle cybersecurity challenges in the real world.

Now, it's time to apply the concepts we've learned. I encourage you to explore further, experiment with Python scripts, and continue learning about cybersecurity principles and techniques.

But our adventure doesn't end here! The next chapter will explore the fascinating world of software testing and game development. Prepare to level up your Python skills and embark on exciting new challenges.

DEBUGGING, TESTING, AND GAME DEVELOPMENT

I n this exciting chapter, we're going to transform into code detectives. Our mission? To master the art of finding and fixing those cunning little bugs that hide in the nooks and crannies of our programs. But worry not; this journey is packed with fun and discovery!

And hold on because there's more! While we're sharpening our detective skills, we'll also be stepping into the shoes of a game creator. That's right – you will design and develop your computer game using Python!

DEBUGGING TECHNIQUES WITH PDB AND IPDB

As a new coder, you might wonder, "What exactly is debugging?" Debugging is like being a detective in a crime drama, where the crime is a bug in your code, and you're the hero who will crack the case.

In computer programming and engineering, debugging is a journey that unfolds in several stages. It begins by recognizing a glitch, followed by the task of pinpointing its origin. Once the root cause is identified, the following steps involve rectifying the issue directly or devising alternative solutions to circumvent it.

Finally, the debugging process thoroughly tests the applied fixes or workarounds to ensure their effectiveness and functionality.

Why is Debugging Important?

- Solve Mysteries in Your Code: Like a detective solves a mystery, debugging helps you solve issues in your code. It's about understanding why your program isn't working as expected.
- Learn from Mistakes: Each bug you find and fix is a learning opportunity. You get to know your code inside out – a crucial step to becoming a great programmer.
- Save Time and Stress: Mastering debugging will save you hours of frustration. Imagine fixing problems in minutes instead of hours!

Meet Your Debugging Sidekicks: pdb and ipdb

pdb (Python Debugger): Think of pdb as your magnifying glass. It lets you pause your program, look at what's happening at each step, and test changes. It's like hitting the pause button on a movie to understand a complicated scene.

Basic Commands:

- list (l): View the current line of code you're inspecting.
- step (s): Move to the following line of code, stepping into function calls.

- next (n): Move to the following line without stepping into function calls.
- break (b): Set a breakpoint to pause your program.

ipdb: Imagine ipdb as your high-tech detective goggles. It does everything pdb does, but it does so with superpowers. It's more user-friendly and lets you write Python commands directly in your debugger.

Why ipdb?

- Easier to read outputs.
- You can use Python's full power to inspect your program.
- It's integrated with IPython, making it more interactive.

Becoming a Code Detective

- Start Small: Begin with simple debugging tasks. Try adding a breakpoint and see what your variables are doing.
- Practice Makes Perfect: The more you debug, the better you'll get. Try to find bugs in your code intentionally and then fix them.
- Keep Calm and Debug On: Sometimes debugging can be frustrating. Remember, every bug you fix makes you a stronger programmer.

THE GREAT DEBUGGER'S GUIDE TO PDB AND IPDB

Meet PDB, your trusty built-in debugger ready to tackle any coding challenge. A hidden gem known as PDB exists in Python programming—the Python Debugger. It's your trusty sidekick in the battle against bugs, boasting a user-friendly command-line interface that simplifies debugging.

PDB performs the job efficiently with its simple command line interface, offering all the essential debugger features you need to troubleshoot your code like a pro.

But why stop there when you can take your debugging experience to the next level? Enter IPDB, the ultimate upgrade for your debugging toolkit. With its friendlier command line interface and seamless integration with IPython, IPDB brings a whole new level of ease and efficiency to your debugging process.

So, how do you get started on this epic debugging adventure? Integrating PDB directly into your code unlocks a world of debugging possibilities. In this small tutorial, we'll show you how to harness the power of Python's built-in debugger to tackle bugs head-on.

Sometimes, spells don't work as expected. That's where our magical tools, pdb and ipdb, come in. They're like your magic wands, helping you peek behind the curtain of your code and fix what's amiss.

Step 1: Summoning PDB- The Classic Wand

Imagine you're walking through a mysterious forest (your code), and you come across a fork in the path (a bug). It would be best to have your map and compass (pdb) to navigate.

Summoning the Magic:

Abracadabra! Add import pdb; pdb.set_trace() in your code where you think the mystery begins.

```python
import pdb; pdb.set_trace()
```

Run your script normally. Your program will pause right where you placed the magic spell.

```python
# Example Python script

def calculate_sum(a, b):
    import pdb; pdb.set_trace()  # Summoning PDB
    return a + b

result = calculate_sum(5, 10)
print(result)
```

Here, pdb.set_trace() is placed inside the calculate_sum function. When you run this script, execution will pause before the return statement.

Navigating the Forest:

- l (list): Reveals the surrounding area of your code.
- n (next): Take a step forward in the code.
- s (step into): Dives deeper into a function.
- c (continue): Keep walking until the next breakpoint or the end of the forest.

```python
# Continuing from the previous snippet...

# PDB is active here
# Try these commands:

l          # List the code around the current line
p a, b     # Print the values of variables 'a' and 'b'
n          # Go to the next line
p result   # Print the value of 'result' (will show an error as 'result'
```

Step 2: Wielding ipdb - The Enhanced Wand

Now, let's add some sparkle! ipdb is like a wand with extra glitter. It's brighter, easier to use, and makes your debugging journey more enjoyable.

Casting the Enhanced Spell:

- Install ipdb: Run pip install ipdb in your terminal (your magic portal).
- Invoke ipdb: Place import ipdb; ipdb.set_trace() in your code where the dragon might be hiding (the bug).

Using Your Sparkly Wand:

- Autocomplete: Just like guessing the end of a spell, ipdb suggests code completions.
- Colorful Output: It's like having a rainbow guide your way, making it easier to read the paths (code).
- Direct Python Commands: You can cast any Python spell (command) directly in ipdb.

Step 3: Mastering the Art

- Set Multiple Breakpoints: Place several magic markers to stop at different points.
- Inspect Variables: Peek into your spell ingredients (variables) to see if they're what you expect.
- Change Variables on the Fly: Like altering ingredients in a potion, you can change variables in real time to test different outcomes.

Step 4: Setting Traps for Bugs

Sometimes, you need to lay traps to catch bugs. In PDB, these are called breakpoints:

- break (or b): Set a breakpoint at a specified line number or function. Your program will pause again when it reaches this point.

```python
# Example Python script

def mystery_function(x):
    for i in range(5):
        x += i
    return x

# Set a breakpoint at line 4
import pdb; pdb.set_trace()
result = mystery_function(10)
```

When you run this script, you can set a breakpoint inside the loop:

```css
b 4, i == 3  # Set a breakpoint at line 4 when i equals 3
```

Step 5: The Art of Interrogation

As a code detective, sometimes you need to interrogate your variables to find the culprit:

- ! (bang): Precede a command with ! to execute it as a Python command.
- locals and globals: Use these to inspect local and global variables

```python
# Continuing from any paused state in PDB

where     # Display the call stack
locals() # Print local variables
!x = 20  # Change the value of 'x' using the bang (!) command
c          # Continue execution until the next breakpoint or end of the program
```

Step 6: Exiting the Scene Gracefully

Once you've gathered your clues and are ready to leave the debugger:

- quit (or q): Gracefully exit PDB and end your debugging session.

```python
# While in PDB

q         # Quit the debugger and terminate the program
```

The Final Trick: Patience and Practice

Remember, even the most significant magicians didn't learn their craft overnight. Be patient with yourself. Every time a spell doesn't work (your code breaks), you're learning.

DEBUGGING MADE EASY: SOLVING BROKEN CODE LIKE A PRO

Suppose we have a Python function that's supposed to calculate the average of numbers in a list, but it's not working as expected:

```python
def calculate_average(numbers):
    total_sum = 0
    for number in numbers:
        total_sum += number
    average = total_sum / len(numbers)
    return average

numbers_list = [2, 4, 6, 8, 10]
print("The average is:", calculate_average(numbers_list))
```

Identifying the Bug

You run this code expecting to see the average of the numbers, but it throws an error: ZeroDivisionError: division by zero. This error message is our first clue – it seems there's an issue when the code tries to divide by the length of the list.

Using PDB to Debug

Let's add a PDB trace to our code to inspect what's going on:

```python
def calculate_average(numbers):
    import pdb; pdb.set_trace()  # Adding PDB trace
    total_sum = 0
    for number in numbers:
        total_sum += number
    average = total_sum / len(numbers)
    return average

numbers_list = []
print("The average is:", calculate_average(numbers_list))
```

Notice that I've changed numbers_list to an empty list, which simulates the scenario that leads to the error.

Step-by-Step Debugging

Run your script. The script will pause at the pdb.set_trace() line.

Inspect the numbers list:

```css
p numbers
```

You'll see it's an empty list [].

Identify the Problem:

- Since the list is empty, len(numbers) returns 0.
- Dividing by zero causes the ZeroDivisionError.

The Solution:

- We need to handle the scenario when the list is empty.
- Let's add a condition to check if the list is empty and return 0 or a message indicating the list is empty.

Fixing the Code:

Update the function as follows:

```python
def calculate_average(numbers):
    if not numbers:  # Check if the list is empty
        return "List is empty, cannot compute average"
    total_sum = 0
    for number in numbers:
        total_sum += number
    average = total_sum / len(numbers)
    return average
```

Test the Fixed Code:

When you run the code with an empty list, it should return "List is empty, cannot compute average" instead of throwing an error.

By stepping through the code with PDB, we could pinpoint the exact location and nature of the bug and then implement a fix. This example demonstrates the power of debugging tools like PDB in efficiently resolving issues in your code. Keep practicing with different scenarios to enhance your debugging skills.

WRITING TESTS USING UNITTEST AND PYTEST

Imagine you're building a spaceship. You'd want to test each part. You wouldn't strap yourself in and blast off without ensuring everything works perfectly.

That's what testing in coding is like. It's about ensuring your software spaceship is ready for a smooth journey.

Let's explore how testing with tools like unittest and pytest is crucial, focusing on seven key benefits:

1. Saves Money

Catching a bug in a spaceship when it's still in the hangar is much cheaper than fixing it in space! Similarly, identifying and fixing issues early in the development process through testing saves money. The longer a bug goes unnoticed, the more expensive it becomes.

2. Security

Software security is like the shields on your spaceship. Tests act as a first line of defense, helping to identify vulnerabilities before they can be exploited. Regular testing ensures the software's shields are always up and ready against potential threats.

3. Quality of the Product

Testing is the polish that makes your spaceship shine. It's about more than just finding bugs; it's about ensuring the software runs smoothly, efficiently, and as expected. High-quality software leads to a better user experience and a more reliable product.

4. Satisfaction of the Customer

Your spaceship passengers (users) expect a safe and comfortable journey. Testing ensures that the software meets or exceeds customer expectations. Happy users are likelier to continue using the product and recommend it to others.

5. Enhancing the Development Process

Testing is like a navigator for your spaceship, guiding the development process. It helps developers understand the codebase better, promotes cleaner, more organized code, and encourages them to write testable, usually more reliable code.

6. Ease of Adding New Features

Imagine adding a new engine to your spaceship. You'd want to ensure it integrates seamlessly with the rest of the ship. Testing makes adding new features effortless, ensuring that new changes don't break existing functionality.

7. Determining the Performance of the Software

Testing your spaceship's speed and maneuverability ensures it performs well in all conditions. Similarly, testing assesses the performance of your software, ensuring it runs fast, efficiently, and reliably under various scenario

TESTING YOUR CODE WITH CONFIDENCE: AN INTRODUCTION TO PYTEST

Chapter 1: The Pillars of Testing with unittest

unittest is like the trusty hammer in your toolkit. It's a part of Python's standard library, ready to help you build robust tests.

A Simple unittest Example:

Let's say we have a function add_numbers that we want to test:

```python
# Our simple function
def add_numbers(a, b):
    return a + b
```

Now, let's write a basic test using unittest:

```python
import unittest

class TestAddNumbers(unittest.TestCase):

    def test_addition(self):
        self.assertEqual(add_numbers(3, 4), 7)

if __name__ == '__main__':
    unittest.main()
```

We create a test case, TestAddNumbers, that inherits from unittest.TestCase. We then write a test method, test_addition, to assert that our add_numbers function gives the correct output.

Chapter 2: The Art of pytest – Sleek and Simple

pytest is like the high-tech drill in your testing toolbox. It's not built into Python but incredibly powerful and user-friendly.

A Basic pytest Test:

First, ensure you have pytest installed:

```bash
pip install pytest
```

Now, let's write a test for the same add_numbers function:

```python
# No need to import pytest, just write the test

def test_addition():
    assert add_numbers(3, 4) == 7
```

Running the Tests

- With unittest, you run the script containing your test case.
- For pytest, navigate to the directory containing your test file and run pytest in your terminal.

THE GREAT CODE TESTING CHALLENGE

Be brave coders, ready for a delightful twist in our coding adventure! It's time to wear your testing hats and dive into a fun, hands-on exercise. If you accept the assignment, your mission is to write a test for a simple yet mysterious function. Are you ready to become a testing maestro? Let's embark on this quest!

Your Objective:

You have a magical function named mystery_function. Your task is to unravel its secrets by writing a test. Here's the function:

```python
def mystery_function(a, b):
    return a * b
```

Seems simple, right? But looks can be deceiving in the world of code!

The Exercise:

Choose Your Tool: Will you pick the trusty unittest or the sleek pytest for your journey? The choice is yours!

Set Up Your Test Environment:

- If you're team unittest, create a new Python file for your test.
- If you're on the pytest squad, start a new Python script.

Write Your Test:

- Think of different scenarios for mystery_function. What if a and b are positive? Negative? What happens with zeros?
- Write assertions to test these cases. For example, if a = 5 and b = 3, the function should return 15.

```python
# Example test case for pytest
def test_mystery_function():
    assert mystery_function(5, 3) == 15
    # Add more assertions here!
```

- For unittest, wrap these assertions in a test method inside a class that inherits from unittest.TestCase.

Run Your Tests:

- For unittest, run your test file as a regular Python script.
- For pytest, run pytest in your terminal in the directory where your script is.

Observe the Results:

- Do all your tests pass?
- What happens if you change the function's behavior? How do the tests react?

Why This Exercise Rocks:

- Discover the Power of Testing: You'll see firsthand how tests can validate the expected behavior of your function.
- Boost Your Debugging Skills: Writing tests is like putting on x-ray glasses that let you see through your code, making bugs easier to spot and fix.
- Unleash Your Creativity: There's no single right way to test. Feel free to experiment with different scenarios!

Reflect on Your Adventure:

Once you've run your tests, take a moment to ponder:

- How did crafting these tests change your understanding of the mystery_function?
- Can you think of other ways to break and catch the function with a test?

Remember, testing is your powerful ally in the grand coding narrative, ensuring every chapter of your code story is flawless. Happy testing, and may your coding journey be ever joyful and bug-free!

CREATING YOUR FIRST GAME WITH PYGAME

In this journey, Pygame is your mystical scroll – a powerful ally that transforms the complex language of Python into a canvas for your creativity. Whether you dream of distant galaxies, enchanted forests, or futuristic cities, Pygame is the key that unlocks these universes.

As you embark on this adventure, you'll learn to mold the virtual environment, orchestrate movements, and narrate stories through interactive gameplay. Each step is a brushstroke in your grand masterpiece, from painting celestial backgrounds to guiding characters with a mouse flick.

Step 1: Preparing for Lift-off – Install Python and Pygame

Before we can soar among the stars, we need to prepare our spaceship:

- Install Python: Make sure you have Python installed on your computer. It's the fuel for our Pygame spaceship.
- Install Pygame: Open your command console and type pip install pygame. This command installs the Pygame modules, the building blocks for our game.

Step 2: Crafting the Cosmos – Implement a Background

Every game needs a captivating backdrop; our digital canvas:

Initialize Pygame and Create a Window:

```python
import pygame
pygame.init()
window = pygame.display.set_mode((800, 600))  # Width and Height of the game window
```

Load and Display the Background:

- Find or create an image for your game's background.
- Load the image in Pygame and draw it onto the window.

```python
background = pygame.image.load('path/to/your/background.jpg')
window.blit(background, (0, 0))
pygame.display.update()
```

Step 3: Setting the Universe in Motion – Scroll the Background

To create a sense of movement, let's make our background scroll:

Animate the Background:

- Continuously shift the background image position to create a scrolling effect.
- Update the display in a loop and handle the event when the game window is closed.

Step 4: Introducing Our Hero – Add a Moving Sprite

Our game needs a star! Let's add a character (sprite) controlled by mouse movement:

Create a Sprite:

- Load an image for your character.
- Draw it on the window, similar to the background.

Make It Follow the Mouse:

- In the game loop, get the mouse position.
- Place the sprite at the mouse coordinates.

Step 5: The Grand Assembly – Put It All Together

Now, let's combine all these elements:

The Game Loop:

- This is where the magic happens. Your game will continuously update the screen, handle user inputs, and render graphics.

Bringing It to Life:

- Implement the scrolling background.
- Add the sprite following the mouse.
- Make sure to update the display and handle events (like closing the window).

Python isn't just about data analysis and web development; it's also a fantastic game development platform. Today, we will harness Python's power to build classic games: a basic version of 'Pong' and 'Snake'. These projects are fun and a great way to deepen your understanding of Python. Let's dive in!

Crafting 'Snake' - A Pythonic Challenge

'Snake' involves a player controlling a line that grows in length, with the line itself being a primary obstacle. The player navigates the snake to consume items, avoiding collision with the walls or the snake's tail.

Step 1: Initial Setup

- Initialize the Game Environment: Set up Pygame and create the main game window.
- Design the Playing Field: Draw the grid where the snake will move.

Step 2: Creating the Snake

- Draw the Snake: Start with a small snake. You can represent it as a list of squares.
- Snake Movement: Program the snake to respond to keypresses to change direction.

Step 3: The Apple

- Place the Apple: Randomly place an apple on the grid, which the snake can eat.
- Growth Mechanism: Each time the snake eats an apple, it grows in length.

Step 4: Game Over Conditions

- Check Collisions: The game ends if the snake hits the wall or itself.
- Restart the Game: Offer the option to restart the game after a game is over.

Building 'Pong' - The Retro Classic

One of the earliest arcade video games, Pong, is a simple table tennis simulation. The player controls a paddle by moving it vertically across the left or right side of the screen. They can compete against another player or the computer.

Step 1: Setting Up

- Initialize Pygame: Start by setting up Pygame, which provides the game's graphical interface.
- Create the Game Window: Set up the window where the game will be played.

Step 2: Paddles and Ball

- Draw Paddles: Use Pygame's drawing functions to create two rectangular paddles.
- Create a Ball: Draw a circular ball and define its movement.

Step 3: Game Mechanics

- Moving the Paddles: Use keyboard input to control the paddles.
- Ball Movement: Program the ball to move across the screen, bouncing off walls and paddles.

Step 4: Scoring and Winning

- Keep Score: Track each time the ball passes a paddle.
- Winning Condition: Set a score that determines the winner.

Putting It All Together

Ready to level up your understanding of the topic? Hit play on the YouTube video below and let the visuals and explanations do the rest. It's an easy way to deepen your knowledge and gain valuable insights in no time

https://www.youtube.com/watch?v=E8fmDDtaHLU

PIXEL PIONEERS: YOUNG CODERS' JOURNEY TO CREATING THEIR FIRST GAMES WITH PYGAME

Andrei is a young visionary who transformed his curiosity into a spectacular game development showcase. His story isn't just about writing lines of code; it's a tale of how he turned pixels into adventures, all thanks to the magic of YouTube tutorials from Clear Code.

Embarking on a Pythonic Quest with Clear Code

In the labyrinth of YouTube, Andrei stumbled upon a channel that would change his digital life forever – Clear Code. This channel wasn't just a collection of tutorials; it was a treasure map to the secrets of game development. With a narrative as compelling as the games they emulated, Clear Code demystified the complex realms of coding, making concepts like loops, functions, and variables as intriguing as the levels of Super Mario, the vast landscapes of Minecraft, or the mystical realms of The Legend of Zelda.

Andrei dove into this learning adventure with his eyes set on the prize. "I chose YouTube for learning because it's free, and the video explanations surpass any text guide," he said, his determination shining through his words.

From Viewer to Creator: Andrei's Gaming Galaxies

Guided by Clear Code, Andrei embarked on his coding crusade, weaving together lines of Python to bring his gaming fantasies to life. He conjured not one but two unnamed masterpieces in the survival action genre. Picture a digital arena where the player battles against relentless foes through swift movements and strategic clicks. Here, every enemy wave conquered, and every level-up is a testament to Andrei's growing prowess in game development.

The Revelation of a Hidden Game Wizard

It was a journey shrouded in mystery, even to his family. Andrei's brother, Gen Sosa, unveiled his sibling's newfound craft. "We thought he was just watching videos; little did we know, he was piecing together his digital realms," Gen expressed, awe coloring his voice. In a mere three weeks, Andrei had mastered the art of game-making and delved deep into the complexities of programming.

Your Turn to Embark on a Coding Adventure

Andrei's story is more than just about coding; it's about the boundless potential within each of us to learn and create. Like Andrei, you too can embark on any journey in game development or any other passion that calls to you. The digital world is your oyster, and resources like Clear Code are your map and compass, guiding you through the seas of learning and creativity.

So, why wait? Grab your keyboard; who knows what incredible games you'll create or the fantastic skills you'll discover along the way? Your adventure awaits!

What's Next: Python Projects and Challenges

As we turn the page to our next chapter, get ready to put all your skills to the test. We're diving into an ocean of exciting Python projects and challenges.

Expect to encounter projects that will stretch your imagination, deepen your understanding of Python, and enhance your problem-solving abilities. From automating daily tasks to analyzing data, the possibilities are endless. Each project is an opportunity to showcase your talents and add impressive work to your coding portfolio.

So gear up, fellow coder! A thrilling adventure awaits, where you'll solidify your coding skills and unleash your creativity. Get ready to showcase your Python prowess to the world. The next chapter is not just about learning; it's about transforming your ideas into reality!

PYTHON PROJECTS AND CHALLENGES

Calling all aspiring Python pioneers! It's time to venture beyond the boundaries of conventional coding and explore the uncharted territory of Python projects. In this chapter, you'll break free from the constraints of tutorials and textbooks as you dive headfirst into hands-on project building.

From simple scripts to sophisticated applications, the sky's the limit. So, buckle up and get ready to unleash your inner Python wizard!

CODING CHRONICLES: PYTHON PROJECTS TO IGNITE YOUR CREATIVITY

Each project cements your understanding of Python and encourages you to sprinkle in your creativity and personal flair. Whether crafting a web application, visualizing data, developing a game, or automating a routine task, a world of possibilities awaits you!

1. Small Web Application:

Project Idea: Create a personal blog or a portfolio website.

Skills Utilized: Flask or Django for the backend, HTML/CSS for the front end.

Customization Tips:

- Integrate a user authentication system.
- Add a feature to post and edit blog entries.
- Implement a comment section for interaction.

2. Data Visualization Project:

Project Idea: Build an interactive dashboard showing real-time data, like stock prices or weather.

Skills Utilized: Python libraries like Pandas for data manipulation, Matplotlib or Seaborn for visualization, and Dash for web integration.

Customization Tips:

- Allow users to select different data sources.
- Create interactive elements like dropdowns to filter data.
- Implement responsive graphs that update with new data.

3. Simple Game:

Project Ideas: Develop classic games like Tic-Tac-Toe, Pong, or Snake.

Skills Utilized: Pygame for game development, basic AI algorithms for game logic

Customization Tips:

- Add different difficulty levels.
- Introduce power-ups or additional features as the game progresses.
- Design unique graphics and sound effects to enhance the gaming experience.

4. Basic Automation Script:

Project Ideas:

- File Organizer: Script that sorts files in a directory into folders based on file type or name.
- Automated Email Sender: A program that sends emails at scheduled times.

Skills Utilized: OS and shutil libraries for file manipulation, and smtplib for email sending.

Customization Tips:

- For the file organizer, add functionality to rename files in a user-defined pattern.
- In the email sender, implement a feature to attach files or customize email templates.

JUMPSTART YOUR PYTHON JOURNEY

The key is to start with small, manageable projects that build on the skills you've learned in each chapter. Let's explore a variety of fun and engaging projects that combine different programming concepts, encouraging creativity and customization.

Here's a guide to help you get started:

1. Mad Libs:

- Skills: Basic input/output, string manipulation.
- Idea: Create an interactive story game.

```python
noun = input("Enter a noun: ")
verb = input("Enter a verb: ")
adjective = input("Enter an adjective: ")
story = f"Today I saw a {noun} which was {verb} in a {adjective} way."
print(story)
```

Tip: Add complexity with more parts of speech and longer stories

2. Guess the Number Game:

- Skills: Looping, conditionals, random modules.
- Idea: The computer picks a random number; the user guesses it.

```python
import random
number = random.randint(1, 100)
guess = None
while guess != number:
    guess = int(input("Guess a number between 1 and 100: "))
    if guess < number:
        print("Higher!")
    elif guess > number:
        print("Lower!")
print("You guessed it!")
```

Tip: Add hints to guide the user.

3. Rock, Paper, Scissors:

- Skills: User input, random module.
- Idea: Play against the computer.

```python
import random
choices = ["rock", "paper", "scissors"]
computer = random.choice(choices)
player = input("Rock, Paper, or Scissors? ")
print(f"Computer chose {computer}")
# Add logic to determine the winner
```

- Tip: Use a dictionary to map choices to what they can beat.

4. Hangman:

- Skills: String manipulation, loops, conditionals.
- Idea: Classic word-guessing game.
- Tip: Start with a predefined list of words and randomly select one for the game.

5. Countdown Timer:

- Skills: Time module, loops.
- Idea: Set a timer and countdown to zero.

```python
import time
seconds = int(input("Enter time in seconds: "))
while seconds:
    mins, secs = divmod(seconds, 60)
    timer = f'{mins:02d}:{secs:02d}'
    print(timer, end="\r")
    time.sleep(1)
    seconds -= 1
print("Time's up!")
```

6. Password Generator:

- Skills: Random module, string manipulation.
- Idea: Generate a random, secure password.

```python
import random
import string
length = int(input("Enter password length: "))
chars = string.ascii_letters + string.digits + string.punctuation
password = ''.join(random.choice(chars) for i in range(length))
print("Your password is:", password)
```

7. QR Code Encoder/Decoder:

- Skills: Working with QR code libraries.
- Project Idea: Create a tool that generates and decodes QR codes.

```python
import qrcode
# QR Code Generation
qr = qrcode.QRCode(version=1, box_size=10, border=5)
qr.add_data('https://example.com')
qr.make(fit=True)
img = qr.make_image(fill='black', back_color='white')
img.save('example_qr.png')
```

Tips:

- Use the QRcode and Pillow libraries to generate and process images.
- Add a GUI using Tkinter for a more interactive experience.

8. Tic-Tac-Toe AI

- Skills Utilized: Basic AI algorithms, game logic, Python.
- Project Idea: Enhance a simple Tic-Tac-Toe game by implementing an AI opponent using the Minimax algorithm.

```python
board = [" " for _ in range(9)]
def print_board():
    row1 = "|".join(board[0:3])
    row2 = "|".join(board[3:6])
    row3 = "|".join(board[6:9])
    print(row1, row2, row3, sep="\n")
```

Tips: Start by creating a 2D array for the board. Research the Minimax algorithm for the AI, which is a perfect fit for such zero-sum games.

9. Binary Search Implementation

- Skills: Algorithms, recursion.
- Project Brief: Implement a binary search function in Python.

```python
def binary_search(arr, target):
    low, high = 0, len(arr) - 1
    while low <= high:
        mid = (low + high) // 2
        if arr[mid] == target:
            return mid
        elif arr[mid] < target:
            low = mid + 1
        else:
            high = mid - 1
    return -1
```

Tips: Understand how to divide the search space in half and recursively search. Test your algorithm with large data sets.

10. Minesweeper

- Skills Used: Multi-dimensional arrays, event handling.
- Project Idea: Develop the Minesweeper game with a grid of cells containing mines.

```python
import random

def create_board(dim_size, num_bombs):
    # Create the board and plant bombs
    board = [[0 for _ in range(dim_size)] for _ in range(dim_size)]
    bombs_planted = 0
    while bombs_planted < num_bombs:
        loc = random.randint(0, dim_size**2 - 1)
        row = loc // dim_size
        col = loc % dim_size
```

```python
        if board[row][col] == -1:
            continue  # Skip if there's already a bomb

        board[row][col] = -1  # Plant a bomb
        bombs_planted += 1

        # Add 1 to neighboring cells
        for r in range(max(0, row-1), min(dim_size-1, row+1)+1):
            for c in range(max(0, col-1), min(dim_size-1, col+1)+1):
                if board[r][c] != -1:
                    board[r][c] += 1
```

```python
        # Add 1 to neighboring cells
        for r in range(max(0, row-1), min(dim_size-1, row+1)+1):
            for c in range(max(0, col-1), min(dim_size-1, col+1)+1):
                if board[r][c] != -1:
                    board[r][c] += 1
    return board

def print_board(board):
    for row in board:
        print(" ".join(str(cell) for cell in row))
```

```python
# Example usage
dim_size = 5
num_bombs = 5
board = create_board(dim_size, num_bombs)
print_board(board)
```

Tip: Start with a text-based version to understand the game's logic before moving on to GUI programming with libraries like Pygame or Tkinter.

11. Sudoku Solver

- Skills Used: Backtracking algorithm.
- Project Idea: A program that can solve Sudoku puzzles.

```python
def is_valid(board, num, pos):
    # Check if the number can be placed in the position
    row, col = pos
    # Check row
    if num in board[row]:
        return False
    # Check column
    if num in [board[i][col] for i in range(9)]:
        return False
    # Check box
    box_x, box_y = row // 3, col // 3
    for i in range(box_x*3, box_x*3 + 3):
        for j in range(box_y*3, box_y*3 + 3):
            if board[i][j] == num:
                return False
```

```python
def solve_sudoku(board):
    empty = find_empty(board)
    if not empty:
        return True  # Puzzle solved
    row, col = empty

    for num in range(1, 10):
        if is_valid(board, num, (row, col)):
            board[row][col] = num

            if solve_sudoku(board):
                return True

            board[row][col] = 0  # Reset the cell
```

```python
        return False  # Trigger backtracking

def find_empty(board):
    for i in range(len(board)):
        for j in range(len(board[0])):
            if board[i][j] == 0:
                return (i, j)
    return None

# Example board (0 represents an empty cell)
board = [
    [5, 3, 0, 0, 7, 0, 0, 0, 0],
    [6, 0, 0, 1, 9, 5, 0, 0, 0],
    ...
]

solve_sudoku(board)
for row in board:
    print(row)
```

Tip: Understand how the backtracking algorithm works. It's a recursive approach that tries filling cells with numbers and back-tracks when it encounters an invalid state.

12. Photo Manipulation in Python

- Skills Used: Working with libraries like PIL or OpenCV.
- Project Idea: Create a script for basic photo editing like resizing, filters, or rotations.

```python
from PIL import Image

# Load an image
img = Image.open('photo.jpg')

# Resize the image
img = img.resize((300, 300))

# Save the edited image
img.save('resized_photo.jpg')
```

Tip: The Python Imaging Library (PIL) is very powerful. Explore its different methods for image manipulation, such as cropping, rotating, and applying filters.

13. Markov Chain Text Composer

- Skills Used: Data structures, probability.
- Project Idea: Generate random, coherent text based on the Markov Chain model.

```python
import random

def build_markov_chain(text, k=3):
    # Build the chain
    pass

def generate_text(chain, size=100):
    # Generate text using the chain
    pass

# Use a sample text to build the chain and then generate new text
```

Tip: Start with a small text and a simple Markov chain (like order 2 or 3) to see how different inputs affect the generated text's coherence.

14. Pong

- Skills Used: Basic game development, GUI programming.
- Project Idea: Create a classic Pong game.

```python
import pygame

# Initialize pygame and set up the game window
pygame.init()
screen = pygame.display.set_mode((800, 600))

# Game loop
while True:
    # Handle events
    for event in pygame.event.get():
        if event.type == pygame.QUIT:
            pygame.quit()
            exit()

    # Update game state

    # Draw everything
    pygame.display.flip()
```

Tip: Break down the game into smaller components, such as the paddle, ball, and scoring system. Implement and test each component separately.

15. Snake

- Skills Used: Keyboard event handling, game logic.
- Project Idea: The classic Snake game where the snake grows as it eats items.

```python
import pygame

# Initialize pygame and set up game window, snake, and food positions
# Game loop
while True:
    # Handle events and snake movement
    # Check for snake colliding with food or itself
    # Update game state and redraw
```

Tip: Focus on getting the snake movement

16. Connect Four

- Skills Used: Array manipulation, win condition checks.
- Project Idea: Develop the Connect Four game where players drop discs into a grid.

```python
def create_board():
    return [[None for _ in range(7)] for _ in range(6)]

def drop_piece(board, row, col, piece):
    board[row][col] = piece

def is_valid_location(board, col):
    # Check if column is not full
    pass

def check_winner(board, piece):
    # Check horizontal, vertical and diagonal win
    pass

# Game initialization and loop
```

Tip: Pay attention to how you check for a winning condition. You'll need to check horizontally, vertically, and diagonally.

17. Tetris

- Skills Used: Complex game logic, collision detection.
- Project Idea: Build the Tetris game with falling geometric shapes.

```python
import random

# Define the shapes and the board
shapes = [...]
board = [...]

def check_valid_position(board, shape, offset):
    # Check if the shape can be placed at the offset position
    pass

def clear_rows(board):
    # Clear completed rows
    pass

# Game loop
while True:
    # Handle piece movement and rotation
    # Check for game over
```

Tip: Managing the different Tetris pieces and their rotations can be challenging. Consider defining each shape and its rotations as a set of coordinates.

18. Online Multiplayer Game

- Skills Used: Network programming, real-time data handling.
- Project Idea: Create a simple multiplayer game where players interact online.

```
# This is a simplified example. You'll need a server script and a client script.
import socket

# Server setup
server = socket.socket(socket.AF_INET, socket.SOCK_STREAM)
server.bind((socket.gethostname(), 1234))
server.listen(5)

# Client setup
client = socket.socket(socket.AF_INET, socket.SOCK_STREAM)
client.connect((socket.gethostname(), 1234))

# Sending and receiving data
# Remember to handle this in both the server and client scripts.
```

Tip: To understand the basics of network communication in gaming, start with a basic text-based or simple graphical game.

19. Web Scraping Program

- Skills Required: HTTP requests, HTML/CSS knowledge, parsing HTML/XML.
- Project Idea: Create a Python script to scrape website data.

```python
import requests
from bs4 import BeautifulSoup

URL = 'http://example.com'
page = requests.get(URL)

soup = BeautifulSoup(page.content, 'html.parser')
# Find elements by ID, class, or tag
```

Tip: Familiarize yourself with the structure of the HTML content you want to scrape. Tools like browser developer tools can be beneficial.

20. Bulk File Renamer

- Skills Required: File I/O, regular expressions, basic understanding of filesystems.
- Project Idea: Develop a program to rename multiple files based on specific patterns or rules in a directory.

```python
import os

def bulk_rename(dir_path, old_ext, new_ext):
    for filename in os.listdir(dir_path):
        if filename.endswith(old_ext):
            new_name = filename.replace(old_ext, new_ext)
            os.rename(os.path.join(dir_path, filename), os.path.join(dir_path, new_

# Example usage: bulk_rename('/path/to/directory', '.txt', '.md')
```

Tip: Always test your script in a safe environment first to avoid accidentally renaming essential files.

21. Weather Program

- Skills Required: Working with APIs, JSON data handling, and basic GUI (optional).
- Project Idea: Write a program that fetches and displays weather information from an API. Users could input a city name, and the program would display the current weather conditions, temperature, or forecast.

```python
import requests

def get_weather(city):
    API_KEY = 'your_api_key'
    URL = f"http://api.openweathermap.org/data/2.5/weather?q={city}&appid={API_KEY}"
    response = requests.get(URL)
    return response.json()

# Example usage
weather_data = get_weather('London')
```

Tip: You must sign up for an API key from a weather data provider like OpenWeatherMap.

22. Code a Discord Bot with Python

- Skills Required: Basic understanding of APIs, asynchronous programming, and Python.
- Project Idea: Create a Discord bot to perform tasks like sending automated messages, responding to user commands, or moderating a Discord server.

```python
import discord
from discord.ext import commands

bot = commands.Bot(command_prefix='!')

@bot.event
async def on_ready():
    print(f'Logged in as {bot.user}')

@bot.command()
async def ping(ctx):
    await ctx.send('pong')

bot.run('your_token_here')
```

Tip: Start with simple commands and gradually add features like responding to events or handling user inputs.

23. Space Invaders Game

- Skills Required: Basic game development, collision detection, GUI programming.
- Project Idea: Develop a clone of the classic Space Invaders game. Your version could include multiple levels, score tracking, and increasing difficulty.

```python
import pygame

# Initialize pygame and set up the game window, player, enemies, and bullets
# Game loop
while True:
    # Handle player movement and shooting
    # Move enemies and handle collisions
    # Update game state and redraw
```

Tip: Break down the game into manageable parts: player movement, enemy behavior, bullet mechanics, and collision detection.

BEGINNER-FRIENDLY PYTHON PROJECT IDEAS WITH CODE SNIPPETS

1. Math Exercises

Skills Required: Basic arithmetic operations, loops, user input/output.

Project Idea: Create a program that generates random math problems (like addition, subtraction, multiplication, division) and asks the user to solve them. It's a great way to practice basic programming concepts and operations.

```python
import random

def generate_math_problem():
    operation = random.choice(['+', '-', '*', '/'])
    num1 = random.randint(1, 10)
    num2 = random.randint(1, 10)
    problem = f"{num1} {operation} {num2}"
    return problem, eval(problem)

problem, answer = generate_math_problem()
user_answer = float(input(f"Solve the problem: {problem} = "))
print("Correct!" if user_answer == answer else "Wrong!")
```

2. Mad Libs Generator

Skills Required: String manipulation, user input/output.

Project Idea: Write a program that takes user input (like nouns, verbs, and adjectives) and inserts it into a pre-made story template. This fun project helps you understand how to work with strings and get user input.

```python
noun = input("Enter a noun: ")
verb = input("Enter a verb: ")
adjective = input("Enter an adjective: ")

story = f"Today I saw a {noun}. It was {adjective} and decided to {verb}!"
print(story)
```

3. Number-Guessing Game

Skills Required: Conditional statements, loops, random number generation.

Project Idea: Create a game where the computer randomly selects a number, and the user has to guess it. After each guess, the computer can give hints like 'too high' or 'too low.' It's a fantastic way to get comfortable with loops and conditions.

```python
import random

secret_number = random.randint(1, 100)
guess = None

while guess != secret_number:
    guess = int(input("Guess a number between 1 and 100: "))
    if guess < secret_number:
        print("Too low!")
    elif guess > secret_number:
        print("Too high!")

print("Congratulations! You guessed it right.")
```

4. Password Generator

Skills Required: Working with strings and randomization.

Project Idea: Develop a program that generates a random, strong password every time it's run. You can experiment with including special characters, numbers, and varying lengths. This project introduces randomness and string manipulation.

```python
import random
import string

def generate_password(length):
    characters = string.ascii_letters + string.digits + string.punctuation
    return ''.join(random.choice(characters) for i in range(length))

password = generate_password(10)
print("Generated password:", password)
```

5. Rock, Paper, Scissors

Skills Required: Basic Python, conditional statements.

Project Idea: A simple implementation of the classic game "Rock, Paper, Scissors" against the computer. It's a great exercise to practice conditional logic.

```python
import random

choices = ["rock", "paper", "scissors"]
computer_choice = random.choice(choices)
user_choice = input("Choose rock, paper, or scissors: ")

if user_choice == computer_choice:
    print("It's a tie!")
elif (user_choice == "rock" and computer_choice == "scissors") or \
     (user_choice == "scissors" and computer_choice == "paper") or \
     (user_choice == "paper" and computer_choice == "rock"):
    print("You win!")
else:
    print("You lose!")
```

6. Hangman

Skills Required: String manipulation, loops, basic data structures.

Project Idea: Code the classic game of Hangman, in which the player guesses a word one letter at a time. This project enhances your skills in handling strings and loops.

```python
import random

words = ['python', 'java', 'kotlin', 'javascript']
chosen_word = random.choice(words)
display = ['_' for _ in chosen_word]

while '_' in display:
    letter = input("Guess a letter: ")
    if letter in chosen_word:
        for i in range(len(chosen_word)):
            if chosen_word[i] == letter:
                display[i] = letter
    print(' '.join(display))

print("You guessed the word!")
```

7. Contacts List

Skills Required: Basic data structures and file I/O (optional).

Project Idea: Create a program to add, remove, modify, and display contacts (like names and phone numbers). You can even try storing this information in a file. This project introduces you to data handling and basic file operations.

```python
contacts = {}

def add_contact(name, phone):
    contacts[name] = phone

def remove_contact(name):
    contacts.pop(name, None)

add_contact("Alice", "123-456-7890")
add_contact("Bob", "987-654-3210")
remove_contact("Alice")

print(contacts)
```

8. Tic-Tac-Toe

Skills Required: Multi-dimensional arrays, basic game logic.

Project Idea: Build a simple Tic-Tac-Toe game that can be played on the console. This is an excellent exercise for understanding arrays and game logic.

```python
board = [" " for _ in range(9)]

def print_board():
    for i in range(3):
        print(board[i*3] + "|" + board[i*3+1] + "|" + board[i*3+2])

def player_move(player):
    position = int(input(f"Player {player}, choose your position (1-9): "))
    board[position-1] = player

# Simplified game loop
for turn in range(9):
    print_board()
    player_move("X" if turn % 2 == 0 else "O")
```

9. Web Scraper

Skills Required: HTTP requests, parsing HTML/XML, basic understanding of web content.

Project Idea: Write a script to extract specific information from websites (like news site headlines). This project will help you learn how to fetch and parse web content.

```python
import requests
from bs4 import BeautifulSoup

URL = 'http://example.com'
page = requests.get(URL)
soup = BeautifulSoup(page.content, 'html.parser')

# Example: Find all paragraphs
for para in soup.find_all('p'):
    print(para.get_text())
```

10. Alarm Clock

Skills Required: Working with time and dates and user input/output.

Project Idea: Develop an alarm clock where you can set a specific time; when it's reached, the program notifies you. This project introduces you to handling dates and times.

```python
import datetime
import time

alarm_time = input("Set alarm time (HH:MM:SS): ")
while True:
    current_time = datetime.datetime.now().strftime("%H:%M:%S")
    if current_time == alarm_time:
        print("Alarm ringing!")
        break
    time.sleep(1)
```

11. Currency Converter

Skills Required: API usage, handling JSON data, basic math operations.

Project Idea: Create a program to convert an amount from one currency to another using real-time exchange rates fetched from an API. It's an excellent way to learn about APIs and JSON.

```python
import requests

def convert_currency(amount, from_currency, to_currency):
    API_KEY = 'your_api_key'
    url = f"https://api.exchangerate-api.com/v4/latest/{from_currency}"
    response = requests.get(url)
    rates = response.json()['rates']
    return
```

12. Automate Social Media Messages

Skills Required: Understanding of APIs and automation concepts.

Project Idea: Write a script to automatically post messages or updates to social media platforms at scheduled times. This project introduces you to social media APIs and the concept of automation.

```python
import tweepy

def send_tweet(tweet):
    consumer_key = 'your_consumer_key'
    consumer_secret = 'your_consumer_secret'
    access_token = 'your_access_token'
    access_token_secret = 'your_access_token_secret'

    auth = tweepy.OAuthHandler(consumer_key, consumer_secret)
    auth.set_access_token(access_token, access_token_secret)

    api = tweepy.API(auth)
    api.update_status(tweet)

# Example Usage
send_tweet("Hello world! This is my first automated tweet.")
```

GUIDED PROJECT IDEAS FOR EACH MAJOR TOPIC

Let's embark on an exciting journey through the world of coding with project ideas that bring each major topic to life. Imagine crafting your creations, from intelligent machines to captivating websites and beyond.

Here's how you can make it happen:

Data Handling: Uncover Hidden Stories in Data

Become a data detective with a project that turns raw data into captivating stories. Imagine you have a treasure trove of weather data. Your mission? Unravel its secrets using Python's Pandas library and bring it to life with striking visualizations using Matplotlib or Seaborn.

The Mission:

- Gather Your Tools: Arm yourself with pandas and matplotlib. pyplot.
- Discover the Data: Load a dataset, perhaps a CSV filled with intriguing weather patterns.
- Cleanse and Prep: Tidy up your data - handle those pesky missing values and outliers.
- Analyze: Dive deep. What tales do the averages, maximums, and patterns tell?
- Visualize: Transform your findings into a visual feast with graphs and charts.
- Conclusions: Be the Sherlock Holmes of data, deducing insights from your visual storytelling.

Machine Learning: Forge Your AI

Using scikit-learn and a housing dataset, you will create a crystal ball that projects real estate prices into the future.

The Steps:

- Summon Libraries: Gather your allies - numpy, pandas, matplotlib, and scikit-learn.
- Prepare the Battlefield: Load and normalize your housing dataset.
- Training and Testing: Split your data into allies and challengers.
- Create the Oracle: Forge your linear regression model.
- Train Your Model: Teach your oracle the ways of the housing market.
- Test Your Creation: Unleash your model on the test data and marvel at its predictions.

Web Development: Weave Your Web

Imagine crafting a personal portfolio website, a digital canvas to showcase your journey and achievements. Using HTML, CSS, and JavaScript, you're going to build a website that's uniquely yours.

The Blueprint:

- Design Your World: Sketch the layout - a welcoming header, an 'about me' story, a gallery of projects, and a contact portal.
- Craft in HTML: Bring your design to life, section by section.
- Style with CSS: Dress your website in colors, fonts, and styles. Make it pop!

- Animate with JavaScript: Add a sprinkle of interactivity for that extra charm.
- Launch: Send your website into the world, perhaps with GitHub Pages as your vessel.

Database Management: Master the Art of Data Keeping

Enter the world of databases by creating a book inventory application. Your tools? SQL and Python. Your goal? A seamless system to manage the comings and goings of books.

The Journey:

- Blueprint Your Database: Design a schema with tales of authors and adventures.
- Speak SQL: Whisper queries to create your database and its chambers (tables).
- Pythonic Power: Infuse your application with Python - add, find, update, and delete books.
- Interface Magic: Conjure a simple yet powerful command-line interface.
- Test and Triumph: Run your application and watch it manage a library of books seamlessly.

Networking and Security: Connect the World Securely

Embark on a quest to build a basic chat application, a virtual room where thoughts and words travel instantly. Your tools? Python's socket library and a dash of threading magic.

The Path:

- Construct the Server: Create a digital tower that listens to the whispers of the internet.

- Craft the Client: Forge the key that connects to the server's realm.
- Multi-threading Mastery: Juggle multiple conversations with the art of threading.
- Exchange of Words: Weave the functionality of sending and receiving messages.
- Guard Up: Implement basic security measures to keep your conversations safe.

Mobile App Development: Bring Ideas to Fingertips

The Dream: Picture a To-Do List App, a personal assistant living in the smartphone. Using a framework like Flutter or React Native, you'll create an app that organizes life, one task at a time.

The Creation:

- Design for Touch: Sketch a user-friendly interface - a list that holds tasks and buttons that command action.
- Environment Setup: Prepare your workshop with Flutter or React Native.
- Craft the front end: Paint your app with UI components.
- Backend Logic: Give life to your app - let it add, display, and manage tasks.
- Emulate and Test: Watch your app come to life in an emulator or actual device.

Each of these projects is a doorway to a new realm of coding, where you'll learn, create, explore, and, most importantly, have fun. So, grab your keyboard, and let the adventure begin.

Crafting a CNN-Based COVID-19 Detector from Chest X-Rays: A Comprehensive Tutorial

Detecting COVID-19 from chest X-ray images using Convolutional Neural Networks (CNNs) is a fascinating application of deep learning in the medical field. This project involves image processing, machine learning, and data analysis. Here's a breakdown of the tools and techniques and a step-by-step guide to implementing this project.

Tools and Techniques Used

- Python: The primary programming language for the project.
- TensorFlow and Keras: Popular deep learning libraries for building and training the CNN model.
- OpenCV or PIL: Python libraries for image processing tasks.
- NumPy and Pandas: For data manipulation and numerical computations.
- Scikit-learn: For data preprocessing and evaluation metrics.
- Dataset: A collection of chest X-ray images, including labeled images of patients with and without COVID-19 (datasets can be found in online repositories like Kaggle).
- CNN (Convolutional Neural Network): The core machine learning technique for image classification.

Step-by-Step Implementation

Step 1: Prepare the Environment

- Install Python and necessary libraries (tensorflow, keras, opencv-python, numpy, pandas, scikit-learn).

- Obtain a dataset of chest X-ray images.

Step 2: Data Preprocessing

- Load the dataset using Pandas or directly with Keras image processing tools.
- Preprocess the images:
- Resize images to a uniform dimension (e.g., 224x224 pixels).
- Normalize pixel values.
- Augment the data if necessary to avoid overfitting.
- Split the dataset into training, validation, and test sets.

Step 3: Build the CNN Model

- Define the CNN architecture in Keras:
- Use convolutional layers (Conv2D) with activation functions like ReLU.
- Include pooling layers (MaxPooling2D) to reduce dimensionality.
- Use dropout layers (Dropout) to reduce overfitting.
- Flatten the output and use dense layers (Dense) for classification.
- The final layer should have a softmax activation function for multi-class classification.
- Compile the model with an appropriate optimizer (like Adam), loss function (like categorical_crossentropy), and metrics (like accuracy).

Step 4: Train the Model

- Fit the model on the training data using model.fit().
- Use the validation data to adjust parameters and avoid overfitting.

Step 5: Evaluate the Model

- After training, evaluate the model's performance on the test set.
- Use metrics like accuracy, precision, recall, and F1-score.
- Confusion matrices can be helpful to see the model's performance in each class.

Step 6: Model Tuning and Improvement

- Based on the evaluation, adjust model parameters or try different architectures.
- Experiment with different numbers of layers, layer sizes, and hyperparameters.

Step 7: Deployment (Optional)

- Once the model performs satisfactorily, it can be deployed as part of a medical imaging analysis tool or integrated into a web application using frameworks like Flask or Django.

Additional Considerations

- Data Privacy: Ensure that the dataset used complies with privacy laws and ethical standards, especially when dealing with medical data.

- Clinical Validation: Clinical deployment requires extensive validation and regulatory approval.
- Collaboration with Medical Professionals: Work alongside healthcare experts to understand the nuances of the data and the clinical relevance of the model's predictions.

This project is not only a great way to apply deep learning skills but also has the potential to significantly impact healthcare, especially in situations like the COVID-19 pandemic.

INTERACTIVE ELEMENT

Get ready to kickstart your coding voyage with our easy-to-use Project Planning Workbook! It's designed to be your guide, helping you transform your brilliant ideas into reality. Whether you aim to build the next viral app or automate everyday tasks, this workbook is your companion in the journey from conception to completion.

1. Goal Setting: Visioning Your Project

- Project Title: Give your project a catchy and descriptive name.
- Objective: What problem does your project solve, or what value does it add? Be as specific as possible.
- Target Audience: Who will benefit from your project? Understanding your audience helps tailor the project to their needs.
- Success Criteria: Define what success looks like. Is it a fully functional app, a certain number of users, or something else?

2. Design Blueprint: Crafting Your Project's Blueprint

- Sketches/Wireframes: Draw basic layouts of your project's interface or structure. Use paper sketches or digital tools like Balsamiq or Figma.
- Feature List: List all the features you plan to include. Start with the core features and then add optional ones.
- User Flow: Outline the steps a user will take to navigate your project. This can be a flowchart showing different user interactions.

3. Technology Stack: Choosing Your Tools

- Programming Languages: Decide on the primary languages you'll use.
- Frameworks and Libraries: Identify any frameworks or libraries to assist your development.
- Tools and Software: List the tools for designing, coding, and testing your project.
- Platforms: Where will your project live? Is it a web app, mobile app, or desktop software?

4. Step-by-Step Implementation: Breaking Down the Process

- Milestones: Break your project into smaller, manageable milestones. For instance, 'Complete User Login System' or 'Design Home Page'.
- Tasks for Each Milestone: List the specific tasks needed to achieve each milestone.
- Timeline: Assign a realistic timeline to each task and milestone, considering your availability and complexity.

5. Testing and Validation: Ensuring Quality

- Testing Plan: How will you test each part of your project? Consider unit tests, integration tests, and user testing.
- Feedback Mechanisms: Plan how you'll gather feedback, whether from beta testers, surveys, or analytics tools.

6. Launch and Post-Launch Plan

- Deployment Strategy: How will you deploy your project? Will it be a phased rollout or all at once?
- Marketing Plan: If applicable, how will you market your project? Consider social media, blogs, or community forums.
- Maintenance Plan: How will you handle updates and bug fixes post-launch?

7. Reflection and Learning

- Challenges and Solutions: Document any significant challenges you faced and how you resolved them.
- Learnings: Reflect on what you've learned throughout the process.
- Future Improvements: Consider potential improvements or additional features for future versions.

Completing this Project Planning Workbook gives you a comprehensive roadmap from a mere idea to a fully realized project. This structured approach keeps your project organized and enhances your learning experience, ensuring that each step is a valuable lesson in your coding journey.

SEGUE

And that's a wrap on this exciting chapter of your coding story! Think of it as a treasure chest you've just filled with shiny, new coding gems. From crafting your first Python project to mastering the art of turning lines of code into functional tools, you've done more than learn – you've created and achieved.

But don't let the adventure end here. These projects are your playground. Experiment with them, add your personal touch, and watch them evolve. Every modification, every new feature you add, is another chapter in your coding journey.

Next, we'll delve into the world beyond coding—we'll look at the methodologies, vibrant communities, and various pathways you can embark upon as you progress in your Python journey. Get ready to learn how to transform your coding abilities into an enduring journey filled with continuous learning and exploration.

BEYOND THE CODE

Your journey with Python has prepared you for this moment —to transform your skills into superpowers. This chapter is more than just a guide; it's a launchpad for coding superheroes. Learn to enhance your abilities, stay at the forefront of technology, and propel your projects into the spotlight. The era of the coding superhero is here, and you're leading the charge!

BEST PRACTICES FOR YOUNG CODERS

Coding ethics, the significance of writing clean and readable code, and the value of staying curious are foundational principles for every coder, regardless of their experience level.

Let's dive into each of these aspects:

Coding Ethics:

Ethical considerations in coding encompass a broad range of principles to ensure that the software developed is functional but also responsible and equitable. This involves:

- Respecting Privacy: Coders must prioritize user privacy by handling sensitive data securely and transparently.
- Avoiding Bias: It is crucial to develop algorithms and models free from biases and discrimination, as biased software can perpetuate societal inequalities.
- Ensuring Security: Ethical coders prioritize building secure systems to protect users from cyber threats and data breaches.

Clean and Readable Code:

Writing clean and readable code is essential for several reasons:

- Maintainability: Code that is well-organized and easy to understand is more straightforward to maintain and update, saving time and effort in the long run.
- Collaboration: Clear and readable code facilitates collaboration among team members, enabling smoother communication and more efficient teamwork.
- Debugging: Clean code simplifies the debugging process, as it is easier to identify and fix errors in well-structured code.

Staying Curious:

Curiosity is the driving force behind continuous learning and improvement in coding. Here's why it's crucial:

- Learning Opportunities: Curiosity encourages exploration and experimentation, leading to the discovery of new technologies, techniques, and solutions.
- Adaptability: In the rapidly evolving field of technology, staying curious enables coders to adapt to new trends and

developments, ensuring they remain relevant and competitive.

- Innovation: Curious minds are likelier to push boundaries and innovate, creating groundbreaking solutions to complex problems.

NAVIGATING ETHICAL CODING IN THE DIGITAL AGE

The growing impact of software on our daily lives underscores the necessity for ethical coding. Many applications we interact with regularly rely on intricate lines of code, and the decisions made by this code can significantly affect users.

Consider a straightforward example: an AI model designed to assess loan applications using machine learning:

```python
def approve_loan(model, applicant_data):
    return model.predict(applicant_data)
```

This function employs a trained model to evaluate whether a loan application merits approval. However, ethical concerns arise if the model exhibits bias against specific demographic groups, such as race or gender. Biased training data can lead to discriminatory decisions by the model, potentially resulting in the unjust denial of loans to deserving applicants.

THE SIGNIFICANCE OF CLEAN CODE IN SOFTWARE DEVELOPMENT

In the bustling world of software development, the clarity and cleanliness of code stand as the pillars upon which projects rise or fall. Imagine, if you will, a world where startups and tech giants alike have seen their dreams dashed against the rocks of technical debt, all because their code was more tangled than a box of Christmas lights.

It's not just a cautionary tale; it's a reality for those who have deviated too far from the path of clean coding practices.

In his philosophical masterpiece "Twilight of the Idols," Friedrich Nietzsche penned a line that has echoed through the ages: "He who has a why to live can bear almost any how." This nugget of wisdom isn't just for pondering the meaning of life; it's a beacon for anyone navigating the complex seas of software development. When we grasp the "why" behind clean code, we arm ourselves against the "how" of the most daunting challenges.

Breaking old habits can feel like trying to climb a mountain in flip-flops—possible, but why make it so hard on yourself? Convincing someone to shift their ways is a delicate art. It's about showing them a captivating vista they're willing to change course to reach. That's precisely why we're kicking off this journey with a deep dive into the essence of clean code and dispelling myths that cloud our collective coding conscience.

Let's embark on this adventure as coders and as artists and architects of a future where every line of code we write is a stroke of genius, a building block of something enduring and remarkable.

Mastering the Seas of Project Management:

Embarking on a large project can feel like setting sail on a vast ocean. It's easy to drift off course without a map and a compass. But fear not, intrepid explorer!

Here are some tried and true tips to keep your project voyage organized and ensure you reach the shores of success with your sanity intact.

1. Forming Your Fellowship: Every quest begins with assembling a team of heroes. Choose individuals whose skills complement each other to form a well-rounded fellowship ready to tackle any challenge.

2. Charting the Realm: Define the scope of your quest. Know the lay of the land, from the towering mountains (major milestones) to the tiniest pebbles (minor tasks). This clarity will guide your journey.

3. The Timeline Scroll: Carve out your deadlines and milestones on a timeline scroll. These are the key dates when the moons align, marking your path forward through the project's phases.

4. Crafting the Quests: Establish clear goals for every member of your fellowship and the team. Each quest contributes to the success of the more remarkable journey.

5. The Rallying Horn: Set the tone from the outset. A clear, inspiring call to action will echo your team's hearts, driving them forward with purpose and unity.

6. The Messenger Birds: Communicate early, often, and effectively. Let messenger birds fly between team members, carrying news and updates to keep everyone informed and engaged.

7. The Council Meetings: Make every gathering count. Whether it's a war council or a brief rendezvous, ensure that each meeting propels the quest forward, solving problems and strategizing the next steps.

8. The Scroll of Requirements: Gather your requirements like a map to hidden treasure. Once your team understands clearly, let them set sail, trusting in their expertise to navigate the waters.

9. The Stars to Navigate By: Identify Key Performance Indicators (KPIs) that serve as stars guiding your journey. These measurable beacons ensure you're always on the right path.

10. The Beacon Fires: Keep the beacon fires burning, ensuring every team member is updated and aligned. In the darkness, these fires will guide your team home.

11. The Arsenal of Success: Equip your team with everything they need to succeed. From tools to training, ensure they're well-armed to face the challenges ahead.

12. The Shield Against Peril: Anticipate risks and stand ready to address issues swiftly. A strong shield can protect your project from unexpected attacks.

13. The Forge of Improvement: Test, refine, test again. Your project will be tempered and strengthened in the forge, becoming unbreakable.

14. The Bard's Songs: Recognize your team's hard work and achievements. Let the bards sing songs of their courage, boosting morale and fostering a sense of accomplishment.

15. The Map Revisited: Regularly evaluate your progress against the map. Adjust your course as needed, learning from the journey to navigate even more effectively in the future.

With these strategies in your adventurer's pack, you can lead your team through the treacherous yet rewarding terrain of extensive project management. Here's to your success and the epic tales that will be told of your project's journey!

SMOOTH SAILING: 9 PROJECT MANAGEMENT TIPS FOR SUCCESS

Dive into the Project Pool with Management Software

- Put on your flippers and dive into project management software! It's like having a trusty underwater map to help you navigate your tasks and deadlines.

Craft Your Project Blueprint

- Every masterpiece starts with a plan, right? So grab your tools and craft that project blueprint like a seasoned architect, outlining every detail of your project journey.

Set Sail with Your Project Schedule

- Hoist the sails and set the course according to your project schedule! Like a seasoned captain, plot your journey, setting waypoints and milestones.

Race Against Time with Deadlines

- Ready, set, go! Embrace the thrill of the race against time as you work with deadlines, channeling your inner speed demon to meet every project milestone.

Prioritize Like a Pro

- Channel your inner life coach and define your priorities! Think of it as creating your custom playlist – arrange those tasks in order of importance and hit play!

Master the Art of Communication

- Open up the channels and master the art of communication! It's like hosting your talk show, ensuring everyone's on the same page and ready to rock and roll.

Ride the Digital Waves with Kanban Boards

- Hang ten and ride the digital waves with kanban boards! Picture yourself surfing through tasks, effortlessly gliding from "to-do" to "done" with every stroke.

Measure Your Progress with Precision

- Break out your measuring tape and gauge your progress with precision! It's like tracking your fitness goals, celebrating every inch closer to project success.

Embrace Agility and Flexibility

- Be nimble, be quick – embrace agility and flexibility like a seasoned gymnast! It's about being ready to pivot and adapt to any project curveball that comes your way.

RESOURCES FOR ONGOING LEARNING AND GROWTH

Online Learning Platforms with Multilingual Support

- **Coursera:** Offers courses from universities and colleges worldwide, many available in multiple languages, providing a global perspective on Python programming.
- **edX:** Similar to Coursera, edX offers a variety of Python courses from institutions worldwide. Some courses are available in different languages, catering to a global audience.
- **Udacity:** Known for its tech-focused courses, Udacity offers Python programming nanodegrees that include real-world projects, mentor support, and career services.

Books

- "Python Crash Course" by Eric Matthes: A hands-on, project-based introduction to Python for beginners.
- "Automate the Boring Stuff with Python" by Al Sweigart: Focuses on practical programming for total beginners and teaches Python to automate everyday tasks.
- "Invent Your Own Computer Games with Python" by Al Sweigart: Ideal for young learners interested in game development and programming.
- "Python for Kids: A Playful Introduction to Programming" by Jason R. Briggs: Makes learning Python fun with engaging examples and exercises.
- "Learning Python" by Mark Lutz: This is a more in-depth exploration of Python, suitable for those who have grasped the basics and wish to delve deeper.

Communities

- **Stack Overflow:** A vast community of developers where you can ask questions, share knowledge, and learn from others' experiences.
- **Reddit's r/learnpython** is a friendly and helpful subreddit focused on learning Python. It's perfect for asking questions and sharing projects.
- **GitHub:** Explore open-source projects, contribute, and collaborate with other developers. This is a great way to learn by doing and see real-world Python applications.
- **Discord Python Community:** Join Python-related Discord servers where enthusiasts and experts gather to discuss ideas, projects, and challenges.
- **PyCon:** The annual Python conference is a great way to meet other Python enthusiasts, learn from experts, and stay updated on the latest in Python development.

Online Forums and Q&A Sites

- **Quora is a** Q&A platform where you can ask questions in many languages and receive answers from the global community, including topics on Python programming.
- **Stack Overflow:** Catering to non-English speakers, these versions provide a platform for asking and answering programming questions, including Python, in these languages.

Conferences and Workshops

- **EuroPython:** Europe's leading conference for the Python community, offering talks, training sessions, and networking opportunities.
- **PyCon Africa:** Brings together the African Python community to discuss and promote the use of Python across the continent.
- **PyCon APAC:** The Asia-Pacific edition of PyCon, focusing on the Python community in Asia-Pacific countries.

THE ESSENTIAL JOURNEY OF CONTINUOUS LEARNING IN TECH

In the ever-evolving landscape of technology, the compass that guides professionals toward success isn't just skill or talent—it's the insatiable hunger for continuous learning. Imagine technology as an uncharted galaxy, where each star represents a new tool, language, or framework.

The only way to navigate this vast expanse is through relentless exploration and learning, a journey that expands our universe and reveals its infinite possibilities.

Embarking on the Voyage of Continuous Learning

Continuous learning is akin to keeping your spaceship tuned, fueled, and ready for the next adventure. It's an expansive concept transcending traditional classroom boundaries, including a kaleidoscope of formal and informal, structured and unstructured activities.

Whether enrolling in a formal course, shadowing a sage-like senior developer, enthusiastically tackling an alien topic, or simply exchanging knowledge over coffee, every moment is an opportunity to learn.

Adapting to the Technological Cosmos

As new technologies emerge at the speed of light, staying stationary is akin to moving backward. The ability to adapt is crucial, and continuous learning is the engine of this adaptability. It enables professionals to harness the latest advancements, from artificial intelligence to quantum computing, turning these once-novel concepts into tools that solve today's challenges.

Navigating the Needs of Learners

In the educational sphere, continuous learning is the bridge that connects learners with the ever-changing demands of the tech world. It tailors the learning experience to meet students where they are, transforming education from a static journey into a dynamic adventure.

Ensuring Relevance in the Galactic Marketplace

In the competitive tech industry, staying relevant is not just about keeping up—it's about leading the charge. Continuous learning propels professionals and businesses ahead of the curve, turning them into pioneers rather than followers.

Cultivating a Culture of Lifelong Exploration

At its core, continuous learning is about nurturing a mindset of curiosity and exploration—a mindset that sees every challenge as a new frontier, every failure as a lesson, and every success as a stepping stone.

How Challenges and Hackathons Sculpt the Tech Titans of Tomorrow

1. The Crucible of Skill Development: Imagine coding challenges as the ultimate training grounds, where every task tackled sharpens your coding prowess, turning theoretical knowledge into practical expertise.

2. The Key to Interview Invincibility: Hackathons and coding competitions are the battlegrounds where you prepare for the ultimate challenge—acing technical interviews. They're not just tests; they're rehearsals for the moment you'll shine.

3. From Theory to Triumph: Abstract concepts meet real-world problems in these platforms. It's one thing to understand a programming concept; it's another to apply it to create solutions that dazzle and deliver.

4. The Odyssey of Ongoing Education: Learning never ceases in the realm of coding challenges. Each problem is a new lesson, ensuring that your journey of growth and mastery knows no bounds.

5. The Arena of Competitive Programming: Pursue excellence in the electrifying world of competitive programming, where every victory is a testament to your dedication, skill, and strategic prowess.

6. A Universe of Problem Diversity: With an ever-expanding array of challenges, these platforms are a cosmos of creativity, beckoning you to explore new territories and expand your problem-solving horizon.

7. Mastery at Your Fingertips: In this digital age, the path to becoming a programming champion is just a click away. With accessible platforms, the only barrier to entry is your resolve to learn and conquer.

8. Crafting Your Legend: Every challenge conquered and hackathon dominated is a story of triumph, a shining addition to your resume that builds an impressive portfolio of real-world problem-solving prowess.

9. The Edge of Excellence: Standing out is paramount in the crowded tech field. These competitions are your chance to rise above the masses, showcasing your skills in a way that captivates and commands attention.

10. The Euphoria of Problem-Solving: Beyond the accolades and achievements, the true joy of coding challenges lies in the thrill of solving—the exhilarating rush of cracking a problem that once seemed impossible.

UNLEASH YOUR POTENTIAL: WHY CODING CHALLENGES MATTER

Enhanced Problem-Solving Abilities: Engaging in coding challenges introduces you to diverse problems that demand creativity and logical thinking for resolution. Regular practice enhances problem-solving skills, empowering you to approach real-world coding dilemmas confidently.

Coding Proficiency Boost: Through consistent practice, your coding proficiency undergoes refinement. You'll grow more adept at navigating the syntax of your preferred programming language and gain fluency in handling common data structures and algorithms.

Algorithmic Mindset Development: Coding challenges often require optimization for efficiency, prompting you to think algorithmically and devise optimal solutions.

Effective Time Management: Many coding challenges imposed time constraints, necessitating the development of practical time-management skills. This proficiency becomes invaluable, particularly during technical interviews or working on projects with strict deadlines.

Technical Interview Preparation: Coding challenges serve as invaluable preparation for those eyeing careers in the tech industry. They closely simulate the problems encountered in technical interviews, ensuring you're well-prepared for these assessments.

Diverse Skill Set Expansion: Coding challenges span various topics, offering opportunities to explore multiple programming domains—from data structures and algorithms to web development and machine learning.

A COMPREHENSIVE GUIDE TO BUILDING A STANDOUT CODING PORTFOLIO

Here's a detailed blueprint for constructing this castle, ensuring it stands tall and proud in the digital kingdom.

Foundational Stones: Building Your Portfolio

Selecting Your Domain Name

Your domain name is the flag of your castle, visible from afar and beckoning visitors. Choose a name that's memorable and reflective of your professional identity. A combination of your name and your craft (like janedoe.codes or johnsmith.dev) can be both professional and distinctive.

Design and Layout: The Architecture of Your Castle

The design of your portfolio is akin to the architecture of a castle. It should be inviting and impressive, easy to navigate, and reflect your style. Utilize platforms such as GitHub Pages, Wix, or Squarespace for customizable templates to build your site with elegance and efficiency.

"About Me" Section: The Story Behind the Knight

This section is the heart of your castle, where you share the tale of your journey, your passion for coding, and the dreams you chase. Be authentic and engaging, letting your personality shine through. This narrative connects you with your audience personally, making your portfolio memorable.

Featured Projects: Displaying Your Treasures

Your projects are the treasures within your castle, each a testament to your skill and creativity. Highlight a mix of projects that demonstrate your range and depth. For each project, include:

- A compelling title
- A succinct description outlining the challenge, your solution, and the technologies used
- Visuals or demos to bring the project to life
- Links to the live project and its source code on GitHub

Ensure your projects are well-documented, making it easy for visitors to understand and appreciate your work.

Contact Information: Opening the Gates

Make it easy for fellow adventurers, collaborators, and potential employers to contact you. Include your professional email address and links to your LinkedIn, GitHub, or other professional social networks. A contact form embedded on your site can also be a convenient way for visitors to reach out directly.

Fortifying Your Castle: Making Your Portfolio Stand Out

Polish Your Code: The Shine on Your Armor

Ensure that the code behind your projects is as clean and polished as the projects themselves. Well-structured, commented, and clean code reflects your professionalism and attention to detail.

Add Explainers: The Scrolls of Knowledge

For each project, include detailed explainers or README files that guide visitors through the project's journey—the problem, your approach, and the outcome. This not only showcases your technical expertise but also your ability to communicate complex ideas.

Choosing a Strong Domain Name: Your Banner in the Digital Realm

A firm, memorable domain name reinforces your brand and makes it easier for visitors to find and remember your portfolio. Consider a domain name that reflects your niche or personal brand within the tech world.

Aesthetic Appeal: The Majesty of Your Castle

Invest time in making your portfolio visually appealing. Aesthetics matter in the digital realm. A well-designed portfolio, with a harmonious color scheme and intuitive layout, can significantly affect how your work is perceived.

The Power of a Portfolio: Unlocking Opportunities and Showcasing Skills

Demonstrating Skills and Creativity

- A portfolio showcases your coding abilities, problem-solving skills, and creativity, offering potential employers or clients a preview of your capabilities.

Showcasing Expertise

- It highlights your expertise in specific technologies or programming languages, positioning you as an expert in your chosen field.

Standing Out in the Job Market

- In a competitive job market, a strong portfolio differentiates you from other candidates, making you a more attractive option to employers.

Building Confidence

- Creating and maintaining a portfolio fosters a sense of accomplishment and pride in your work, boosting your confidence as a developer.

Fostering Continuous Growth

- A portfolio is a dynamic platform that encourages ongoing learning and improvement, demonstrating your dedication to your craft and adaptability.

MAXIMIZING COMMUNITY POWER: A BLUEPRINT FOR THRIVING IN CODING FORUMS AND CLUBS

As coding technologies and communities expand daily, many newcomers struggle to get their questions answered and resolve issues or bugs.

Unlock Benefits: Step into the vibrant universe of coding forums and clubs where skill growth, networking, and support on projects await.

Organize Engaging Events: Transform engagement with exciting hackathons, coding challenges, and expert talks, turning learning into a celebration.

Share Knowledge Resources: Equip members with tutorials and guides, unveiling the path to advanced coding mastery.

Foster Inclusivity: Create a nurturing space where questions are treasures, and every contribution fuels collective growth.

Celebrate Achievements: Highlight the victories and contributions within the community, inspiring participation and elevating morale.

CRAFTING YOUR CODING NARRATIVE: INTERACTIVE PORTFOLIO PLANNER

```
                                          ┌─── Showcase best work
                        1. Personal Projects ─┤
                                          └─── Highlight used technologies

                                          ┌─── Programming Languages
                        2. Skills ─────────┤
                                          └─── Frameworks & Libraries

                                          ┌─── List previous roles
Portfolio Planner ──────3. Experience ─────┤
                                          └─── Open Source Contributions

                                          ┌─── Relevant Certifications
                        4. Education ──────┤
                                          └─── Online Courses & Workshops

                                          ┌─── Short Biography
                        5. About Me ───────┤
                                          └─── Personal Interests & Hobbies
```

Personal Projects: Showcase your creative and problem-solving skills by describing your projects and technologies.

Skills: List your technical skills, categorized by proficiency and relevance to your projects.

Experience: Detail relevant work experience, internships, or freelance projects, highlighting your contributions.

Education: Include your degree, major/minor, coursework, and certifications.

About Me: Introduce yourself, share your passion for coding, career goals, and what drives you in tech.

Key Takeaways:

- Python is a versatile and powerful tool with endless possibilities.
- Practice and persistence are vital to mastering Python and becoming proficient in coding.
- Building projects is an effective way to solidify your learning and showcase your skills.
- Remember the importance of clean code, ethical coding practices, and staying organized.

Moving Forward:

Now that you've reached this milestone continue your coding journey with enthusiasm and determination. Put the ideas presented into practice by working on new projects, refining your skills, and sharing your knowledge with others.

CONCLUSION

Python stands not just as a programming language but as a beacon of opportunity, igniting the sparks of creativity, innovation, and empowerment across the globe. This book is a tribute to Python's simplicity, a celebration of its flexibility, and an acknowledgment of its immense power to break barriers and democratize technology for all. Through the lens of young trailblazers and seasoned experts alike, we've traveled across a landscape where code becomes the canvas and imagination the limit.

Embarking on this journey, we begin by demystifying the process of Python installation on diverse operating systems, ensuring that the threshold to enter the programming world is as welcoming as possible. We unravel the intricacies of Python's data types and operations, laying a robust foundation that is both comprehensive and approachable.

From setting up Python on any device to unraveling the mysteries of data types, operations, and beyond, we've laid a foundation that is as solid as it is accessible. We've seen how Python serves as the sculptor's chisel, carving out intricate patterns in the vast stone of data and turning raw numbers into compelling stories through visualizations that speak louder than words. With each chart, graph, and plot, a new insight emerges, crafted by the powerful tools nestled within libraries like Matplotlib, Seaborn, Plotly, and Bokeh.

Our expedition through the multifaceted domains of Python—spanning the web's vast landscapes with development frameworks like Flask and Django, streamlining the mundane with automation, unraveling the complexities of artificial intelligence and machine learning with TensorFlow and PyTorch, and fortifying our digital domains against cyber threats—has been a testament to Python's dynamic adaptability.

Each chapter in this journey has illuminated Python's chameleon-like nature, showcasing its ability to seamlessly integrate and evolve within the ever-shifting paradigms of our digital epoch. The tales of TensorFlow breathing life into algorithmic thoughts, Django architecting virtual realms, and Python scripts acting as vigilant sentinels guarding our cyber sanctuaries collectively underscore a singular truth: Python is the language of the future here today.

As our path winds to its close, we're reminded that mastery comes not from the final destination but from the journey itself. The story of a young coder who rose from humble beginnings to innovate with Python is a testament to the transformative power of coding. Their success is not just in the code they wrote but in the barriers they shattered.

So, to you, dear reader, the torch is passed. Keep coding, not as a task, but as a journey of discovery. Let each line of code you write be a step toward mastery. Build, share, inspire, and let the cycle of creativity and learning spin ever onwards. And as you forge ahead, carving your path with Python, remember to share your story, for it could light the way for others just embarking on their journey.

But as with any epic saga, the essence lies not in the destination but in the voyage itself. Our narrative arc—from novices staring wide-eyed at the starlit sky of coding to becoming architects of our digital universes—illustrates the profound transformation that occurs when we engage with Python. The young coder, once a mere apprentice to the art, now stands as a beacon of innovation, a reminder that the potential to reshape our world lies within each of us.

So, as we close this book, let it not be an end but a beacon signaling the commencement of your expedition. Coding is a journey of discovery, an exploration of the boundless frontiers of technology and imagination. Let each line of code you inscribe be a testament to your journey, a building block in the monument of your achievements. Please share your story, for in the sharing, we extend the map of our collective journey, guiding those who follow in our footsteps.

If this guide has been your astral chart through the Python universe, lighting your way and inspiring you to reach beyond, then I urge you to leave a star in the cosmic ledger with your review. Share the tale of your voyage, the wonders you've discovered, and the dreams you dare to dream.

Together, let's ensure the flame of curiosity and the thirst for knowledge burns ever brighter, illuminating the cosmos for the adventurers who will navigate these stars in the coming epochs.

Keep coding, for in the vast expanse of the digital cosmos, your journey is just beginning. May your path through the realms of Python be ever upward, ever onward, and infinitely wondrous.

REFERENCES

https://blog.ktbyte.com/successful-programmers-started-coding-at-a-young-age/

https://www.python.org/downloads/macos/

https://kinsta.com/knowledgebase/install-python/

https://www.geeksforgeeks.org/data-visualisation-in-python-using-matplotlib-and-seaborn/

https://medium.com/@jsonmez/why-learn-python-61621ce49010

https://www.stationx.net/python-for-cyber-security/

http://crufti.com/program-all-the-things-how-to-develop-iot-devices-using-micropython/

https://realpython.com/micropython/

https://realpython.com/pytest-python-testing/

https://www.smartparenting.com.ph/parenting/tweens-teens/two-grade-8-students-create-their-own-video-games-a00391-20220818-lfrm

https://www.dataquest.io/blog/python-projects-for-beginners/

https://www.freecodecamp.org/news/python-projects-for-beginners/

https://www.linkedin.com/pulse/importance-code-quality-why-clean-matters-addant-systems/

https://www.workast.com/blog/9-must-know-hacks-for-effective-project-management-in-2023/

https://www.myhatchpad.com/insight/the-ultimate-guide-to-building-a-coding-portfolio/

https://anywhere.epam.com/en/blog/programmer-portfolio

www.ingramcontent.com/pod-product-compliance
Lightning Source LLC
Chambersburg PA
CBHW071557210326
41597CB00019B/3288